Naturalists' Handbooks 3

Solitary wasps

PETER F. YEO

SARAH A. CORBET

With illustrations by Anthony J. Hopkins

Pelagic Publishing
www.pelagicpublishing.com

Published by Pelagic Publishing
www.pelagicpublishing.com
PO Box 725, Exeter, EX1 9QU, UK

Solitary wasps
Naturalists' Handbooks 3
(2nd edition)

Series editors
S. A. Corbet and R. H. L. Disney

ISBN 978-1-78427-033-9

Digital reprint edition of:
ISBN 0-85546-295-7 (1995) Paperback
ISBN 0-85546-296-5 (1995) Hardback

British Library Cataloguing in Publication Data
A catalogue record for this book is available from the British
Library.

Contents

Editors' preface to the second edition

Solitary wasps fascinate naturalists because they are diverse, often colourful, lively creatures which are absorbing to watch. It is not necessary to have specialised biological training in order to make a valuable contribution to knowledge of the ecology and distribution of these insects, and we hope this book will help to attract more people to study their natural history.

Since the first edition of this book was written, there have been substantial advances in research on solitary wasps, and the text has been revised to take these into account. In particular, much more is now known about the ecology of several genera including two of the three selected as focal examples in chapter 2, *Ammophila* (Field, 1989, 1992) and *Cerceris* (Willmer, 1985a,b). Also, several species have been added to the British list. Of those previously well known on the Continent, all the species that have first appeared in Britain in the twentieth century, beginning with *Microdynerus exilis* in 1937, have been found first in the extreme south east and have then spread north and west. None of them is a soil nester. They all nest in wood or other materials which may have been transported across the Channel.

As well as covering the latest arrivals, the keys now include the species apparently extinct in Britain, to help continental users of the book. The authors and editors thank Jeremy Field, George Else and Geoff Allen for their kindness in helping with the incorporation of this fresh material.

SAC
RHLD
July 1994

Acknowledgements

We are grateful to the late Professor O.W. Richards, and to G.R. Else, M.C. Day, Professor P.S. Corbet, J. Calvert and Dr P.G. Willmer, whose constructive comments on parts of the text have helped greatly. We thank the authorities of the British Museum (Natural History) and the Cambridge University Museum of Zoology for allowing us to study their collections. The late G.M. Spooner lent his unpublished key to the Eumenidae. This help made a vital contribution to the writing of this book, though we, of course, bear full responsiblity for the end result. We also acknowledge our good fortune in being able to include the work of Tony Hopkins, who combines an expert knowledge of wasps with his artistic skill. Some of the wasps were originally drawn for Professor Richards' handbook; we are grateful for the opportunity to re-use them. We thank Dr Anne Bebbington and AIDGAP for suggesting ways of making the keys easier to use.

1 Introduction

Solitary wasps are common insects although they often go unnoticed. They are well-worth looking at; their beautifully elaborate behaviour poses questions of great ecological and evolutionary concern and much remains to be discovered about the biology of even the commonest British species. This book aims to show how interesting solitary wasps can be and give enough information about their natural history to enable readers to make biological observations that will make a real contribution to knowledge; it helps with the naming of wasps and it offers some methods and techniques, references and addresses that may be useful in pursuing the investigation and in communicating the findings.

What are solitary wasps?

The aggressive black-and-yellow wasps familiar to most people are social insects: females co-operate with their sisters and their mother in the maintenance of a colony that may contain hundreds or even thousands of workers as well as a queen. The solitary wasps, on the other hand, do not collaborate in this way. Generally, each female makes a nest of some sort for her own young. She may nest close to others of her species, but except in rare cases which are of special interest, she works alone.

All the wasps, together with bees, ants, sawflies, ichneumon flies and others, belong to the order Hymenoptera, which includes some of the most lively and active of insects. The Hymenoptera can be distinguished from other insects by the way their two pairs of wings link together. A row of microscopic hooks on the hindwing catches on a ridge at the back of the forewing (fig. 1). This coupling arrangement can be demonstrated by gently moving the forewing over the hindwing until the hooks catch on the ridge and the wings zip neatly together.

Each group within the Hymenoptera has its own characteristic lifestyle, behaviour and structure. Fig. 2 shows how the different lifestyles are distributed within the order. (The corresponding morphological differences, which enable us to identify members of these groups without needing to know all about their biology, are given in chapter 3.)

Fig. 1. Hooks linking forewing to hindwing in Hymenoptera.

The wasps are among the most sophisticated of Hymenoptera. The larvae of the Symphyta (sawflies and wood-wasps) generally have legs and biting mouthparts, and often live on plants. But in the group to which the wasps belong, the Apocrita, the larvae are soft white legless grubs with feeble mouthparts. Their survival depends on their mother's ability to lay her eggs in a habitat that will shelter the helpless larvae. In the Parasitica this larval protection is provided by the host, usually an insect or a plant. But most of the Aculeata provide more elaborate

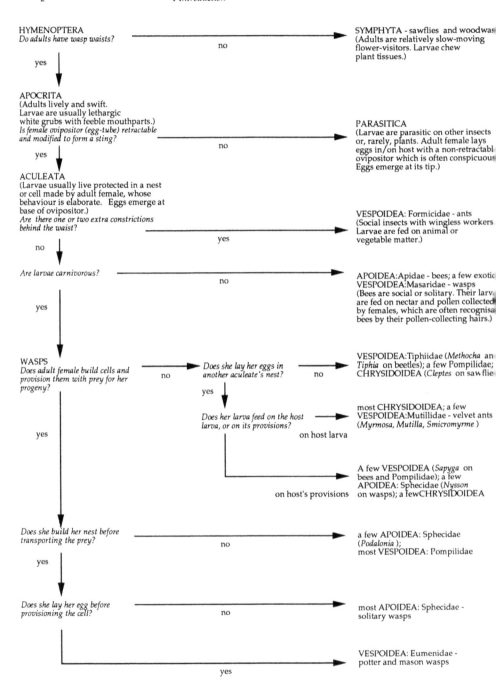

Fig. 2. Scheme illustrating the biological features underlying the (incomplete) classification of the Hymenoptera shown on p. 3 (partly based on Gauld and Bolton, 1988). This scheme should not be used as a means of identification.

protection by constructing and provisioning nests in which they lay their eggs. Among the aculeates, the bees feed their young on the pollen and nectar of flowers; the ants include carnivores as well as herbivores; but the solitary wasps belong to the remaining groups of superfamilies in which the larvae are generally carnivorous. Thus the special skills of adult wasps can be related to provision for their larvae: elaborate selection of an egg-laying site; construction of a nest in soil or wood; and capture of insects or spiders as food for their young.

Another speciality of some hymenopterans, sociality, carries with it a further range of behavioural adaptation; all the ants, some of the bees and some of the wasps are social. Other books in this series will cover some of the social forms. This one limits itself to the British solitary wasps within the families Sphecidae (superfamily Apoidea), Tiphiidae, Mutillidae, Sapygidae and Eumenidae (superfamily Vespoidea) (Gauld and Bolton, 1988)*. It does not cover the spider-hunting wasps, the Pompilidae (superfamily Vespoidea), which are dealt with by Day (1988). The solitary wasps are a mixed assemblage, representing several superfamilies and united by behavioural, rather than morphological, features. It is not easy to give a useful definition of the group. If you find a lively, busy insect that looks something like any of the coloured illustrations (pls. 1–4), you can check that it belongs to one of the groups we deal with by taking it through key I (p. 26). It can then be identified in the other keys.

Order Hymenoptera
  Suborder Symphyta
  (sawflies and woodwasps)
  Suborder Apocrita
    Series Parasitica
    (parasitic Hymenoptera)
    Series Aculeata
      Superfamily Chrysidoidea
        Family Chrysididae
        (ruby-tailed wasps and others)
      Superfamily Vespoidea
        Family Tiphiidae
        (solitary wasps)
        Family Mutillidae
        (velvet ants)
        Family Sapygidae
        Family Formicidae
        (ants)
        Family Pompilidae
        (spider-hunting wasps)
        Family Eumenidae
        (potter and mason wasps)
        Family Vespidae
        (social wasps)
      Superfamily Apoidea
        Family Sphecidae
        (solitary wasps)
        Family Apidae
        (bees)

Finding solitary wasps

To find solitary wasps it is necessary to look in the right place at the right time. Suitable places to search are nest sites or feeding sites. Nests are commonly built in sunny situations. Many wasps nest in holes in dead wood, and a dead tree or branch or post riddled with old beetle-burrows may contain several species, and may be a very rewarding place to watch. Beetle holes in which solitary wasps have nested can sometimes be recognised because they are plugged by the female wasp when she has finished stocking the last cell. Cut or broken stems, particularly of bramble, are also good hunting grounds, and those in which a wasp has nested are recognisable by the hole she has bored down the pith. Some wasps nest in holes in mortar in old walls. Others nest in soil, usually in open sandy areas, or vertical banks of light soil or sand. The holes may be numerous, and on bare open ground the spoil-heaps that some species leave beside their holes may make the nests very conspicuous. Solitary bees nest in similar situations and their nest aggregations are hard to distinguish from those of wasps when there are no insects about.

*References cited under the authors' names in the text appear in full in Further Reading on p. 61.

The best place for making behavioural observations is a nest site with many occupied burrows. An observer sitting quietly beside such a site in warm sunny weather may see the female wasps making their nests or returning from hunting trips carrying prey; or male wasps seeking females and mating with them; or parasites exploring the holes, perhaps entering them to lay an egg or deposit a larva, or trailing a wasp that is provisioning a nest. To see a wasp hunting prey is a rarer privilege. It is not easy to locate the hunting ground.

The adult wasps can sometimes be found taking nectar on such flowers as hogweed or wild carrot.

Wasps are summer insects, and each species has its characteristic season (Richards, 1980). They generally fly only in sunshine, and their activity depends very much on weather and time of day. Even at a populous nesting site wasps are rarely seen out of their nests except during the warmest hours of a sunny day. Except in very hot weather, when they work longer hours, they usually emerge in the late morning and retire in the late afternoon. The disturbance caused by an observer's arrival may make them retreat down their holes, to reappear and eventually emerge some minutes after the disturbance has stopped. It is often necessary to sit quietly for a few minutes before one sees the first sign of a wasp's presence: a face at the entrance of a burrow.

Naming solitary wasps

If an investigator is working in complete isolation, an invented name is perfectly adequate to describe a species of wasp. But if he wants to draw on the pool of information already available on wasps, or contribute observations to that pool, the species must be identified. It is important not to place too much confidence in an uncertain identification. A wasp cannot be identified with certainty by matching the specimen against a picture. Pictures may show what it is likely to be, but for critical identification the wasp must be taken through at least one key and then confirmatory characters must be checked.

Our main keys (pp. 26–56) demand very careful examination of the wasp, which must therefore be dead or anaesthetised. To identify a wasp with certainty it is thus always necessary to take a specimen, and it is the ecologist's abiding dilemma that if he catches an animal to identify it he can no longer watch it; if he leaves it so that he can watch its behaviour he cannot name it with certainty; and if he doesn't know what it is, he doesn't know what to look out for. The Guessing Guide (p. 28) may help to resolve this dilemma in some cases, although it will always be necessary to catch and identify a few specimens eventually.

The Quick-Check Key (pp. 24–25) will make it possible to allocate a wasp roughly to genus, and this knowledge will give the reader confidence and a sense of direction in his progress through the main key.

These keys take account of all the British species of
Sphecidae, Tiphiidae, Mutillidae, Sapygidae and
Eumenidae, but in a few very difficult groups it is not
practicable to distinguish every species in this short book.
Those who wish to go further should consult an expert or a
more advanced publication. The standard work for the
identification of British solitary wasps is by Richards (1980).
We hope that our more elementary key will encourage
serious workers to embark on Richards' key.

Conservation and collecting

It is necessary to collect a few individuals for
identification, but it is best not to collect too many
individuals of the same species at any one locality. Wasps
often nest in aggregations and show site fidelity, each female
returning to nest at the site where she herself developed.
Persistence of the local population may depend on the
maintenance of established sites, some of which may be
hundreds of years old, and removal of entire nest sites can
be very serious indeed. Changes in the landscape, involving
nest site destruction, have already been responsible for a
decrease in the abundance and diversity of solitary wasps in
Britain (Falk, 1991). Much more research is needed on nest
site selection and constancy in various species. Careful
monitoring of all the known solitary wasp nesting
aggregations in a number of areas (for instance, around
particular schools) over a period of years would help us to
assess the likely impact on wasp numbers of collecting and
environmental modification.
        The review of the status of our rare and threatened
species of wasps and bees by Falk (1991) provides a firm
basis for conservation management and a valuable pointer
towards the species for which more information is most
urgently needed.

Distribution in Britain

Warm sandy areas have the largest numbers of
individuals and of species of solitary wasps. Particularly
rich areas are the New Forest, and sandy heathland such as
the Surrey heaths and the Breckland in East Anglia. But the
distribution of individual species remains incompletely
known. Readers of this book are invited to help remedy this
situation by joining the Bees, Wasps and Ants Recording
Society. Details are given in chapter 4.

2 Natural history

Why study wasps?

The behaviour of wasps is consistent enough and rich enough in detail to give abundant material for comparison between species. These insects therefore offer a rare opportunity for exploring the evolutionary development of behaviour. As Evans (1963) has shown, there is sometimes a good match between evolutionary trees based on morphological comparisons, and those based on comparative studies of the inherited components of behaviour, the patterns that are common to the individuals within a species, genus or larger group. Much is known about the morphological interrelationships of wasps because an insect's morphology is enshrined in its cuticle, which is easily preserved and studied in museums. Behaviour, on the other hand, is ephemeral, and its description requires careful and informed observation of living animals in the field. There is still much that can be contributed by careful studies of the behaviour of even the commoner species of solitary wasps in this country. Perhaps readers of this book will undertake such studies. This chapter aims to provide enough background to show which details of behaviour are likely to be significant and worth watching for.

Fig. 3. *Ammophila pubescens* with prey (after Baerends, 1941).

The general pattern of a female solitary wasp's adult life involves preparing a nest consisting of several cells. In each cell she puts prey (fig. 3) as food for her offspring and lays an egg. When an egg hatches the resulting wasp larva eats the prey stored for it in its cell, pupates, and eventually emerges as an adult wasp which makes its way out of the nest, mates, and, if it is a female, begins to prepare and provision a nest for its own eggs. Most solitary wasps show part or all of this pattern, but there is great variation from one species to another in the sequence and the details. To introduce the wasps we describe the adult behaviour of three representatives: *Passaloecus*, a small black wasp that nests in wood and hunts aphids; *Cerceris arenaria*, a larger black-and-yellow sand-nesting wasp that hunts weevils; and *Ammophila pubescens*, a slender black-and-red wasp that nests in sand and hunts caterpillars.

Passaloecus

Wasps of the genus *Passaloecus* (pl. 6.1), being small and black, are easily overlooked although several species are widespread in this country. Since they often inhabit the woodwork of old buildings and fences, they are among the wasps most likely to occur in gardens. The several species of *Passaloecus* look so much alike that they can be distinguished only with a lens or microscope. The observations described here were made on wasps that had been caught, lightly anaesthetised (technique, p. 58), identified, marked with tiny dots of paint (technique, p. 58) so that they could be recognised individually, and then allowed to recover and released again. There were three species of *Passaloecus* living in the same garden: *P. insignis*, *P. gracilis* and *P. corniger* (Corbet & Backhouse, 1975). The first and second of these are conventional aphid-hunters. The third species, *P. corniger*, proved unexpectedly different, as we shall see.

Mating in *Passaloecus* involves a 'sun-dance', which we saw on a sunny sward of honeysuckle leaves one June day. A few females of *P. gracilis* had settled on the leaves in the sun, and numerous males were flying up and down, facing the leaves and a few centimetres out from them, occasionally pouncing on a settled female (or on some other insect, or even on a mark on a leaf). Like so many insect mating assemblies, this sun-dance seems to be strictly localised in space and time. It occurs at the beginning of the flying season, and the males die soon afterwards.

Having mated, females of the aphid-hunting species can be seen house-hunting, walking quickly over the surface of old wood, investigating holes with their antennae. Sometimes a female darts head first into a promising beetle-boring. Usually she emerges within a few seconds and continues her exploration. Occasionally a hole receives more prolonged attention, and eventually she may back into it, a sign that she has selected it as a nest. She then prepares the hole, removing debris from it and sometimes even widening it. She may spend long periods inside, coming to the surface from time to time dusted with fresh sawdust which she clears from her body by vibrating her wings with a short buzz. Sometimes one can hear a similar buzz from a wasp working in the depths of the hole.

When the nest is ready, the female goes hunting for aphids. Many wasps are remarkably quick at finding and catching their prey, and a female that has found a dense colony of aphids may take only one or two minutes to catch an aphid and carry it to her hole. We saw a female *P. insignis* visiting an aphid colony on the lower surface of a leaf of Giant hogweed. She walked up to the colony, and then, with her feet firmly planted on the leaf, she swayed her body and head from side to side repeatedly, as if taking a visual fix on an individual aphid a few millimetres away. Very suddenly, she leapt forward and seized the aphid, and then flew with it back to her nest. Further work may well show a more complicated picture. Unconfirmed observations on a patch of Broad beans infested with Black bean aphid (*Aphis fabae*) suggest that species of *Passaloecus* may sometimes attack aphids directly from the air, and may even become involved in fights with the ants that guard the bean aphids.

The wasp squeezes her captured aphid around the neck region with her mandibles, and after this treatment the aphids are paralysed; they cannot walk, but they remain fresh, with their hearts beating, and dissection may show larvae of parasitic Hymenoptera feeding and moving inside them. The wasp flies back to her nest with her aphid held below her body, gripped by her mandibles around its neck. She lands on the wood a few centimetres from her nest, runs head first into the hole and deposits the aphid. Within seconds she leaves again for the next hunting trip. Occasionally wasps paralyse their prey incompletely, so that the aphids may recover and escape from the nest while the wasp is out. When there are about 20 or 30 aphids in the hole, the wasp lays an egg on one of them and builds a transverse partition, sealing off a cell containing one egg and enough aphids to feed the larva that will hatch from it. She then provisions a second cell above the first, continuing in this way until the nest hole is nearly filled with a row of perhaps three to five cells (fig. 4). Finally she cements over the hole with a plug of pine resin, making repeated journeys to collect it and returning from each with a gobbet of clear resin held in her mandibles.

It is obvious from her behaviour that this material is inconveniently sticky to manipulate, but it offers double protection to her nest, providing a chemical barrier as well as a physical one. Important components of pine resin are the pinenes (fig. 5), terpenoids which repel various insects and so are regarded as a chemical defence whose primary function is the protection of pine trees from attack by herbivorous insects. By using pinene-containing resin to seal her hole, *Passaloecus* is adopting the pine tree's chemical defence. The wasps sometimes camouflage the pale resin by covering it with dark particles.

The above description applies in most particulars to *P. insignis* and *P. gracilis*, but *Passaloecus corniger* is very different in an important respect: instead of hunting aphids for itself, it takes them, one by one, from the nests of *P. insignis*, *P. gracilis*, and even other *P. corniger*. This sort of nest robbery may sometimes go unrecognised because it is hard to detect unless the wasps are marked individually. Some other wasps steal prey from their conspecific neighbours occasionally, but *P. corniger* evidently does so habitually: indeed we have no evidence that it ever hunts for itself.

The hole that a female of the thieving species, *P. corniger*, chooses as her nest may be an untenanted beetle-boring or it may be another wasp's nest that she has just emptied by stealing the aphids from it. When her nest is ready she walks rapidly about, exploring with her antennae and sometimes briefly entering a hole. If she finds a nest being stocked by another wasp she may dart in while the owner is out, and emerge with an aphid which she carries to her own nest. Such robbery is repeated many times in quick succession. If the owner returns there is sometimes a vigorous fight during which each wasp jabs at the other with her sting. Occasionally *P. corniger* re-opens the newly sealed nest of another wasp, laboriously removing droplets of sticky resin from the front door and depositing them as a characteristic ring of white beads around the hole. One by one she carries the aphids to her own nest; presumably she destroys the original owner's egg.

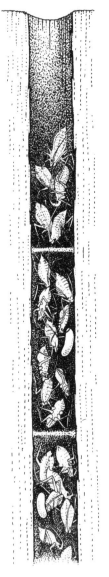

Fig. 4. *Passaloecus* nest in wood, opened to show contents.

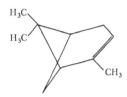

Fig. 5. Molecular structure of α-pinene (after Eisner, 1970).

Because *P. corniger* steals from nests stocked by her own species as well as others, she sometimes receives aphids that have already been stolen more than once. A detailed study of a small area showed that both nest holes and stores may change ownership several times during a busy season. A wasp runs a serious risk of losing her stores, and this puts a premium on their defence. A female *Passaloecus* provisioning her cell often spends much of her time in or near the hole, moving her head alertly, or pursuing and even sometimes attacking intruding wasps.

When the nest is completed and the owner has left it to start another, the safety of its contents depends on a burglar-proof door. The resin plug, though time-consuming for a thief to remove, is not impregnable, and it is possible that the owner may deter intruders by scent-marking. When the wasp larva has hatched and begun to feed, its aphids are not worth stealing, but for the two or three days before this, while her stores remain vulnerable, the owner revisits her sealed nest periodically, trailing the tip of her abdomen around the entrance as if depositing scent. On the only two occasions when we observed *P. corniger* robbing a sealed nest, the owner had been prevented from guarding or scent-marking, once because she was anaesthetised for identification, and once because she drowned in a cup of coffee. Perhaps the scent-mark warns potential intruders that an aggressive owner is not far away. This hypothesis is based on very tenuous evidence and needs further investigation. It is becoming increasingly clear that bees and wasps have a rich vocabulary of chemical signals. Observers watching wasps with this in mind may well be able to develop ideas about the role of chemical communication, and perhaps test these ideas experimentally.

Cerceris

Cerceris arenaria (pl. 8.3), one of the sturdy black-and-yellow wasp-like wasps, is much larger and more conspicuous than *Passaloecus* and correspondingly easier to observe. It is the commonest of the six British species of *Cerceris*, and like the others it nests in sandy soil. Usually many wasps nest in the same area, forming colonies which may include hundreds of nests and extend over many square metres. During the wasps' flying season a colony is a remarkable sight. Each nest, a hole surrounded by a conical mound of excavated soil (fig. 6), looks like a little volcano, and the whole aggregation gives the impression of a miniature lunar landscape. A colony like this is a strong clue that *Cerceris* is present, but to be certain about this it is necessary to check which species are working in it. Other species of solitary wasps and bees make nest-aggregations of this sort too, and three or four species may nest close together.

Fig. 6. Nest of *Cerceris arenaria.*

Colonies of *C. arenaria* are made more obvious early in the season by the presence of many male wasps flying to and fro about 25 cm above the sand, evidently seeking

females. Every now and then a male dives down and
grapples with a female, and some of these encounters result
in mating. It is usual among solitary wasps for males to
mate with newly emerged females, waiting for them to
struggle out from the cells where they developed and
pouncing on them, sometimes before their cuticle has
hardened, or even making feeble digging movements in the
sand as if to help the females out. It is therefore curious that
males of C. arenaria often dive onto females which are
already provisioning their cells, and so presumably must
have mated before. It remains to be seen whether such
females always reject the males, or whether repeated mating
is usual in this species, as it is in an American species of
Oxybelus (Bohart and Marsh, 1960). The mating strategies of
our solitary wasps offer a promising field for research,
particularly in view of the finding that males may help to
protect the nests from parasites (see p. 16) (Alcock and
others, 1978; Peckham, 1977; Thornhill and Alcock, 1983).

When colonising a new site, a female of C. arenaria
must dig herself a fresh burrow in the sand, but in an
established colony she is more likely to save time by digging
fresh cells as side-branches from an existing burrow, such as
the one from which she herself first emerged as an adult
(Willmer, 1985b). There are also records of individuals
usurping the burrows of neighbouring Cerceris or even
Ammophila (Olberg, 1959). As the nest-digging female
penetrates below the loose dry surface sand, moister soil
begins to appear on her spoil-heap, pushed up by the tip of
the wasp's abdomen as she backs up her hole. The finished
burrow may be 5–30 cm deep, branching near the bottom,
and with a cell stocked with about four to twelve weevils at
the end of each of several branches. The wasp probably
makes a larger cell, stocked with more prey, for her female
offspring than for her male offspring (Willmer, 1985b).

The energetic cost to the female of burrowing so
deep is repaid in a microclimatic advantage to her progeny.
At the surface of the unshaded bare sandy soil where
Cerceris nests, microclimatic conditions are desert-like. By
day the sand gets very hot in the sun; by night, loss of
radiation to a clear sky can cool the surface very fast. Deeper
down, temperature changes are less severe. Fig. 7 shows a
temperature profile measured by pushing a small
thermocouple (see p. 59) down a burrow of C. arenaria on a
sunny day. The cells in which Cerceris larvae develop are
deep enough in the soil to avoid extremes of temperature
(Willmer, 1985a), and to experience a steady humidity,
escaping the alternate desiccation and flooding that must
take place at the surface.

The adult wasps, too, are very much at the mercy of
microclimate in the extreme habitat where they choose to
nest. Their activity is confined to warm sunny weather. Both
sexes of C. arenaria normally spend the night in burrows, the
males in special short 'night burrows' and the females in
their nests. In the early morning or during cloudy periods a
female habitually sits at the mouth of her burrow, with her

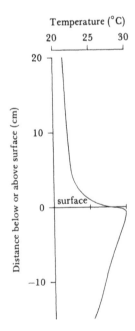

Fig. 7. Temperature profile
down a burrow of Cerceris
arenaria, and in the air above it.

face at the entrance, in the region that will warm up first in the sun. Before taking off she may emerge and bask for a few minutes, as if raising her body temperature to that necessary for flight. Observations on *Passaloecus* indicate a relationship between activity and temperature – wasp temperature (the temperature that would be achieved by a black wasp sitting where *Passaloecus* sits). Willmer (1985a) used fine thermocouples to explore the relationship between temperature and wasp activity much more clearly in the microclimatically more extreme habitat of *Cerceris*. By recording wasp activity at different temperatures she was able to identify the minimum temperature for activity, and to conclude that very high temperatures at the hottest times of day could sometimes prevent the larger individual wasps from working.

A hunting wasp that finds an aggregation of prey can economise on searching time by revisiting it again and again and bringing a fresh prey animal to her nest every few minutes. *Passaloecus insignis* exploits colonies of aphids; the fly-hunting wasp *Mellinus arvensis* (pl. 3.1) waits at dung to ambush flies that come there; *Cerceris rybyensis* (pl. 3.2) haunts the nesting colonies of solitary bees and pounces on the bees as they come in carrying pollen; and the American species *Cerceris blakei* catches weevils from feeding aggregations on flowers. The weevils taken by *C. arenaria* are sometimes abundant enough to be forestry pests, and *C. arenaria* has proved to be a useful biological control agent in Russia (Lomholdt, 1975). But if its weevils are scarce, every capture must involve a long search. When a female goes hunting it may be half an hour or more before she flies back to her nest, using her mandibles and her middle legs to carry, ventral-side-up below her body, a weevil which she has paralysed by stinging it (fig. 8).

Fig. 8. *Cerceris arenaria* with a weevil (after Olberg, 1959).

It is impressive to see a female wasp, with a weevil almost her own size slung below her, fly in low over the colony and hover for a moment in front of her nest before flying down and darting into her hole, still clutching her prey. *C. arenaria* enters her hole unusually quickly; many other wasps land nearby and change their grip on the prey or re-open the temporarily-closed hole before going in. A female *C. arenaria* that is forced to put down her weevil, perhaps to clear away sand that has blocked her hole, will often leave that weevil where it lies and go off to hunt another. A busy *Cerceris* colony may become littered with abandoned weevils. If the wasps were to retrieve this temporarily undefended prey they might risk importing with it the eggs or larvae of parasites.

The cells of *C. arenaria* are sometimes parasitised by ruby-tailed wasps (Chrysididae), and by satellite flies (Sarcophagidae), some of which are viviparous, giving birth to first-stage larvae instead of laying eggs.

Ammophila

Very different from the stocky black-and-yellow *Cerceris* are the slender black-and-red wasps of the genus *Ammophila* (pl. 7.4), perhaps our most handsome solitary wasps. They have received more attention from biologists than the rest, and the story that emerges is a remarkable one.

In Britain we have two species of *Ammophila* and two of the closely related genus *Podalonia* (pl. 2.1). They all nest in sandy areas, often where pine trees grow, and two or three species may occur together. The morphological differences between them are small, but the behavioural differences are surprisingly great. They all nest in holes in the sand and stock their cells with caterpillars. *Ammophila* seeks caterpillars that live on plants, frequently taking species such as the Pine looper *Bupalus piniaria* and the Pine beauty *Panolis flammea*, which, when abundant enough, can both be serious pests of pine trees. *Podalonia*, on the other hand, often hunts the soil-dwelling caterpillars of noctuid moths, known to gardeners as cutworms, which feed on the below-ground parts of plants.

As well as the difference in hunting technique between *Ammophila* and *Podalonia*, corresponding with this difference in the habitat of their prey, there is an important difference in the sequence of their activities. In the British species of *Ammophila*, as in most of the sphecids, a female prepares her nest first and then catches prey and brings it to the nest. *Podalonia*, on the other hand, resembles the more primitive spider-hunting pompilid wasps in that she sometimes catches the prey first. She carries it on foot to a suitable site and may lodge it in a tuft of grass or a forked stem while she digs her nest nearby. The commoner of our two species of *Ammophila*, *A. sabulosa*, often stocks each cell with a single caterpillar, which, because it is necessarily larger than she is, she transports mainly on foot, carrying it along held just off the ground (Field, 1992). But *A. pubescens* provisions each cell with several (usually five to ten) caterpillars, usually small enough for her to fly with them at least part of the way to the nest, and she adopts a strategy not known in any other British solitary species. Instead of sealing the entire ration of food for her larva in the cell with the egg (mass provisioning), she provides only one or two caterpillars initially, and then returns to inspect the nest a day or two later, bringing more caterpillars when her larva is ready to eat them. This behaviour is known as progressive provisioning.

A female of *A. pubescens* may have two or three nests on the go at once, all at different stages of development, and their maintenance depends on remarkable feats of memory and versatility of behaviour. This was revealed by a beautifully detailed observational and experimental study by Baerends (1941). The original paper is in German; Tinbergen (1958) gives an account of these findings in English.

The males of *A. pubescens* emerge a few days before the females and fly about over the nest site waiting for

Fig. 9. *Ammophila pubescens* mating (after Baerends, 1941).

Fig. 10. *Ammophila pubescens* stinging a caterpillar (after Baerends, 1941).

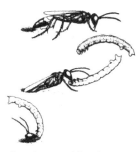

Fig. 11. *Ammophila pubescens* dragging a caterpillar into her nest (after Baerends, 1941).

females to emerge from the sand. When a male sees a freshly emerged female he dives on her, seizing her around the neck with his mandibles. He keeps hold of her and the pair walk or fly about together, mating briefly (fig. 9), for a minute or so at a time, and repeatedly, perhaps ten times in the course of an hour. Soon after mating the female begins to search for a nest site, moving in a characteristic way, alternating short low flights with brief bouts of running. From time to time she pauses briefly to scuff at the sand with her front legs or bite at it with her mandibles. Eventually she spends long enough in one place to make a shallow pit. She may dig several such pits before finally settling on one and deepening it to make a complete nest. She loosens the sand by biting at it and then, holding a load between her forelegs and her head, she flies off and deposits it a few centimetres from the nest. By scattering each load in a different place she avoids the accumulation of a spoil-heap which might act as a clue to searching parasites. The completed nest consists of a vertical burrow about 2 cm long ending in a single oval cell.

Before leaving, the wasp blocks the entrance of her new nest, using lumps of sand, stones or bits of wood which she appears to select with care, picking several up in her mandibles and discarding them before accepting one and carrying it to the nest. Later, when the nest contains caterpillars and a young wasp, it will be covered more carefully; sand will be scuffed over the stones, making it almost invisible.

A. pubescens hunts caterpillars on the heather plants within a few metres of her nest. When she finds one she paralyses it by stinging it ventrally several times (fig. 10) and kneading it with her mandibles. Grasping it behind the head with her mandibles, and supporting it with her forelegs, she carries it home, flying with it unless it is very heavy, in which case she transports it on foot.

By following marked wasps, and by experimentally altering the arrangement of objects near the nest, Baerends showed that the remarkable homing ability of these wasps depends on their learning to recognise landmarks in a particular area. Arriving home with prey, a wasp puts it down beside her nest and unblocks the entrance, scuffing away the sand and removing the larger particles in her mandibles. Then, backing into the hole, she uses her mandibles to drag the caterpillar after her into the nest (fig. 11).

She lays an egg on the first caterpillar, typically on the third abdominal segment. Then she leaves the nest, closing it behind her, and may go to work at another nest not far away. A day or two later she returns, unblocks and enters the nest, and then leaves, re-closing her nest as she goes. Baerends' elegant experiments have shown that this is an inspection visit: if the wasp finds that her larva has eaten its caterpillar, she will bring it a few more during the next day or two, but if not she will wait until another inspection visit shows that the caterpillar has been eaten. About a day later the wasp repeats her inspection, and if food is needed she brings in perhaps six or seven caterpillars in quick succession and then finally closes the hole much more thoroughly than she has done before, ramming down soil into the nest entrance with her head (fig. 12), buzzing loudly as she does so. Other species tamp down the plug using a stone held in the mandibles as a hammer.

Unlike *Passaloecus* and *Cerceris, Ammophila* does not spend the night in her nest. Instead, Baerends' marked wasps left their nest site in the evenings and went to a particular piece of heather, where they spent the night close together, each clinging rigidly to a twig with her mandibles (fig. 13). Sleeping aggregations like this are not easy to find, but they have been recorded for some other wasps and bees, and they may turn out to be widespread. They often consist of males, and they sometimes involve more than one species.

Most adult wasps feed on nectar or honeydew and some also lick the body fluids of the prey collected for their larvae. Both males and females of *A. pubescens* take nectar from flowers. Their unusually long tongues enable them to reach the nectar of shallow flowers like Ling (*Calluna vulgaris*) and Rosebay willowherb (*Chamerion angustifolium*), but on deeper ones like Cross-leaved heath (*Erica tetralix*) they can only feed where a short-tongued bumblebee has already gnawed a hole in the corolla to steal the nectar.

Fig. 12. *Ammophila pubescens* packing soil with her head (after Baerends, 1941).

Provision for the young

The three representative wasps that we have described, *Passaloecus, Cerceris* and *Ammophila*, prepare a nest and transport to it the prey that will form the food for their larvae. This is usual among the more advanced hunting wasps, the Sphecidae. But some solitary wasps invest less time and energy in the preparation of the food and habitat for their young. Some wandering wasps in the superfamily Scolioidea do not transport their prey, but paralyse it and lay one or more eggs on it where they found it. The prey is in a protected situation but not necessarily in a nest; *Tiphia* attacks beetle larvae living in the soil. Others attack insect larvae in holes or burrows of their own: *Methocha* paralyses and lays eggs on Tiger beetle larvae, which live in burrows in sandy soil, and *Myrmosa* (pl. 1.2) and the velvet ants (Mutillidae) (pl. 1.1) enter the cells of solitary wasps and bees and lay an egg there. The larva hatching from it will wait for the bee or wasp larva to develop and then consume it. These wandering wasps provide a behavioural link between the true hunting wasps and the parasitic Hymenoptera, such as ichneumon flies, which lay one or more eggs on or in an insect larva, sometimes paralysing it first by stinging it.

An alternative procedure involving no nest-building or prey transport is followed by kleptoparasites, which lay their eggs in the nest-cells of a bee or wasp and whose larvae feed not on the bee or wasp larvae but on the stored provisions. The scolioid *Sapyga* (pl. 7.8) is kleptoparasitic on

Fig. 13. *Ammophila pubescens* asleep (after Baerends, 1941).

solitary bees, so that its larva has a diet of pollen and nectar, unlike other British wasps whose larvae are carnivorous. *Nysson* (pls. 2.3, 2.4) is a sphecid, clearly related to and descended from full-blooded hunting wasps, but instead of making and stocking her own nest a female *Nysson* lays her egg in a nest-cell of *Gorytes* (pls. 2.5, 2.6, 8.4) or *Argogorytes* (pl. 2.7). When it hatches, the *Nysson* larva destroys the *Gorytes* egg and feeds on the froghoppers stored by the mother *Gorytes*. Some other insects which we refer to loosely as 'parasites' of wasps are in fact kleptoparasites. They include the satellite flies such as *Metopia* (fig. 14) and *Senotainia* (family Sarcophagidae).

Fig. 14. A satellite fly, *Metopia argyrocoephala.*

Defending the prey

A paralysed insect is a valuable food source and a female wasp has to contend with a variety of other competitors for it until she has it safely sealed into her nest. If she puts it down outside her nest it may be carried away by ants, or an egg (or larva) may be laid on it by a kleptoparasitic fly such as *Metopia*. When she has taken the prey, with the fly's egg or larva on it, into her nest-cell, the fly larva will destroy the wasp's egg or larva and demolish the prey. Even inside the nest the prey is subject to theft or usurpation by intruders. It may be removed by another wasp of the same genus (*Passaloecus corniger*) or species (*Ammophila sabulosa*; Field, 1989). The prey may be taken over by another wasp of the same species, who replaces the original egg with her own (*Ammophila sabulosa*; Field, 1989). It may have laid on it the egg or larva of a kleptoparasite, that will feed on the stored prey (flies such as *Metopia* and possibly also some ruby-tailed wasps); or the egg of a true parasite, that will feed on the developing wasp larva (most ruby-tailed wasps).

Studies in which the fate of each prey item is identified, by a combination of behavioural observation and analysis of cell contents (Field, 1989; technique, p. 59), can help to quantify the various threats to the survival of the developing wasp. The risk of loss of prey must have been a major factor in the evolution of prey transport and nest-closing behaviour.

Some females of *Ammophila sabulosa* have only one caterpillar per cell and only one cell per nest, and can therefore lay an egg and seal the nest as soon as the caterpillar is in place. But most wasps have more than one prey item per cell (and more than one cell per nest). This has the advantage that the prey is correspondingly smaller, and can therefore be carried in flight, but the disadvantage that a partly stocked cell must be left undefended while the female wasp is out hunting.

Cerceris arenaria leaves its hole open during hunting trips, and the satellite fly *Metopia* sometimes enters and may deposit its larva in partly stocked cells in the owner's absence. Like some American species of *Cerceris*, *C. arenaria* inconveniences kleptoparasitic flies by accumulating several

beetles, embedded in a plug of loose sand deep in the nest
but just outside the empty cell, before transferring them all
to the cell and then sealing it.

Some wasps, like *Ammophila pubescens*, reduce the
risk of intrusion by temporarily blocking the nest entrance
each time they go hunting. Some kleptoparasites can
penetrate such closures, and *Nysson* will even open one and
then re-seal it when she leaves; but temporary closure
inconveniences all kleptoparasites and must deter some
altogether. It also means that the owner herself must spend
time unblocking her hole whenever she comes home, and
her attention is distracted from her prey while she does so. It
is against this background that the evolution of prey-
carrying techniques comes into focus, as Evans (1962) has
shown in an elegant and stimulating paper.

Wasps that carry their prey in their mandibles
(whether or not they support it with their legs as well
during flight) must leave their nests open as *Cerceris
arenaria* does, or else put down the prey to use their
mandibles and front legs to re-open the hole as *Ammophila
pubescens* does. Some wasps whose prey is small enough to
be carried in flight use only their legs (often the middle or
hindlegs) to grasp their prey, and can re-open the nest after
temporary closure without letting go of the prey. The
extreme of this series is represented by *Oxybelus* (pl. 3.7),
which carries its fly impaled on the sting and keeps it there
while re-opening the nest (fig. 15).

Fig. 15. *Oxybelus uniglumis*
holds a fly on her sting while
re-opening her nest (after
Olberg, 1959).

Temporary nest closure is relatively easy for wasps
nesting in level ground, because their excavated soil can be
used to form a plug. No such material is at hand for wasps
nesting in vertical banks or in wood or twigs, and
temporary closure is correspondingly rare among these. The
wood-nesting *Trypoxylon* (pl. 6.2) adopts an antiparasite
strategy that is altogether different. The males of most
wasps play no part in brood care, but those of some
American species of *Trypoxylon* are said to guard the nest
hole while the female is out hunting spiders (Evans &
Eberhard, 1970). Behaviour of this kind has not been
recorded for British species, but the presence at the nest site
of the males of species like *Cerceris arenaria* (or an American
species of *Oxybelus*: Bohart & Marsh, 1960; Peckham, 1977),
buzzing any insect that approaches, could be of selective
advantage in deterring parasites. Observations made with
this possibility in mind could be of great interest. The time
spent by females in apparently fruitless interactions with
insistent males may prove to be compensated for by a
reduced percentage of parasite damage in the cells made
early in the season when males are about (Willmer, 1985b).

Nest construction

Nest-building behaviour, like prey transport, can be
seen as a series of adaptations reducing the risk of theft or
parasitism of the stored prey. But the nest does more: it
protects the wasp's progeny as well as the progeny's food

supply. In terms of the amount of effort they invest in this protection for the next generation, solitary wasps cover a wide range. As we have seen, wandering wasps such as *Tiphia* (pl. 5.7) use no nest, and velvet ants use the nests in which they find their prey. The other solitary wasps prepare nests in one of three ways. Some, like *Passaloecus*, re-use the nests of other insects, perhaps modifying the hole a bit. Some, like *Ammophila*, excavate their own holes. And some, like *Eumenes* the potter wasp (fig. 16), construct nests using building materials which they have carried to the site. In a classic paper on wasp behaviour, written in English delightfully flavoured by translation from the Japanese, Iwata (1942) described the wasps using these three methods as 'renters', 'diggers' and 'builders'. The word 'renters' is perhaps not quite appropriate, because these tenants give no return to their landlord. It might be fairer to describe them as 'squatters'.

Fig. 16. *Eumenes* at her nest (after Olberg, 1959).

Nests may be made in soil or in wood, with corresponding differences in technique. Soil-nesting species generally dig their own holes, showing species-specific differences in the soil type, degree of compaction, and the slope of the surface in which they nest. On the other hand most wood-nesting species are squatters, re-using second-hand holes made by beetles in posts and dead stumps or branches. Some make cavities by removing the soft pith from broken or cut stems such as bramble or elder; and the cut stems of brambles that have been slashed or pruned provide a rich nesting ground for solitary wasps and bees. Danks (1971) describes the structure and habits of many bramble-nesting species and their parasites, which can be conveniently studied if bundles of cut stems are set out as 'trap-nests' (technique, p. 59).

A wasp's nest usually contains more than one cell. Nests made in soil commonly branch so that each cell is at the end of its own side-tunnel, as in *Cerceris*, but the holes taken over by wasps nesting in wood or pith cavities are often straight and unbranched, and in such cases the wasps' cells may be arranged in a row one above the other, as in *Passaloecus* (fig. 4).

Wasps nesting in wood or stems often have horizontal burrows; any excavated debris or sawdust falls clear when it is discharged from the nest entrance, and accumulation of spoil presents no problem. But most soil-nesting species burrow downwards from a horizontal surface, so that excavated soil has to be carried up to the top of the hole. If it is allowed to accumulate there the spoil-heap makes the nest very conspicuous, as in the case of *Cerceris arenaria*. The hole itself seems to be a cue for parasites – one often sees them investigating holes and other marks of similar size providing visual contrast with the background – but a hole surrounded by a tumulus of excavated soil must be even more obvious.

Some wasps make their holes harder to find by dispersing the excavated soil, and the details of behaviour involved in digging out and, in some cases, scattering soil

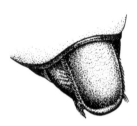

Fig. 17. *Cerceris arenaria:*
pygidium.

Fig. 18. *Mellinus arvensis*
excavating her nest (after
Olberg, 1959).

Fig. 19. *Oxybelus* nest-digging
(after Olberg, 1959).

have been examined and beautifully illustrated with
photographs by Olberg (1959). He recognises four ways of
bringing earth up out of the burrow. Some wasps use two or
three of these. A 'pusher' like *Cerceris arenaria* backs out of
her hole pushing the soil behind her, often using a special
flat area, the pygidium, at the tip of her abdomen (fig. 17),
and forming a mound of soil at the nest entrance (fig. 6). A
'puller' such as *Mellinus arvensis* (pl. 3.1; fig. 18) backs out of
her hole carrying the soil between her head and her front
legs. *Mellinus* leaves the soil at the entrance, but some
pullers walk a few centimetres before putting it down.
'Rakers' excavate and disperse the soil by kicking it
backwards below the body, usually using their specially
modified front legs synchronously (in sphecids) or
alternately (in the spider-hunting pompilids, not included in
this book). *Oxybelus* (fig. 19) looks just like a dog digging for
a bone when she stands on her middle- and hindlegs and
scuffs with her forelegs to send out a spray of sand for
several centimetres behind her body. This scatters the soil so
effectively that there is no spoil-heap to give a clue to the
location of the nest entrance. As Olberg (1959) points out,
when *Ammophila pubescens* scuffs sand away from and over
her closed nest entrance she is providing olfactory as well as
visual camouflage by distributing the *Ammophila*-scented
sand far and wide. 'Carriers' pick up lumps of soil using the
forelegs, the chin and the mandibles and walk or fly away
from the nest with them. *Ammophila pubescens* (fig. 20) uses
this method as well as raking. In fact *Ammophila* and
Podalonia are experts at hiding their nests, raking, carrying,
using the head or a stone held in the mandibles as a tool to
tamp down the plug, and even camouflaging the sealed hole
by dragging twigs or pine needles over it.

Some enemies of wasps' nests must be guided at
least partly by scent, and we have seen how *Passaloecus* may
deter them by using pinenes and scent-marking, and how
Ammophila may confuse them by scattering the sand. Do her
own responses to chemical stimuli help a wasp to discover a
suitable nest site and recognise her own nest? In a series of
experiments on a soil-nesting bee, Butler (1965) showed that
new nests were made only in or near an existing colony,
although apparently suitable habitat remained unoccupied.
But when he moved a lump of soil containing occupied
nests from the colony to a new area, the bees that emerged
from it began to dig nests there, probably in response to the
scent of the transferred soil. In this way he established a new
colony. A chemical cue of this kind is probably important for
solitary wasps, too, but direct evidence for British species
has yet to be obtained. It is well worth looking for because
of its practical importance in managing wasp colonies for
conservation and perhaps even for biological control.

It might be possible to encourage wasps in a colony
threatened by habitat destruction to establish a new nest site
if we knew more about the local distribution of nest sites
and the biology of nest site selection. The study could
involve identifying the common features of a number of nest

Fig. 20. *Ammophila pubescens* carrying sand from her nest (after Olberg, 1959).

sites; watching wasps while they are searching, and comparing the features of acceptable sites at that time with features of the sites that they reject; and examining the consequences of experimental modification of potential nest sites. Females of the American grasshopper-hunting wasp *Sphex ichneumoneus* nest in flat, vegetation-free areas of moderately compacted soil, and Brockmann (1979) showed that their nest-searching is confined to areas that are sunlit and warm at the time.

Prey choice

In their choice of prey solitary wasps are more specific than most predators but less specific than many of the parasitic Hymenoptera. They often confine their attention to a superfamily or order of prey insects. *Passaloecus insignis* takes a variety of aphids; *Cerceris arenaria* takes several species of large weevils; and *Ammophila pubescens* takes a range of caterpillars. Within a species, individual females, even next-door neighbours, may differ from one another in the species of prey they choose, and we must assume that the exact composition of the diet of a larva depends on the luck and experience of its mother when she is hunting.

This prey specificity of solitary wasps enables each species to develop handling techniques appropriate for the prey, and many of these are consistent within a species, or even a genus or subfamily, of wasps. A female that brings back a fresh weevil or aphid every minute or two is clearly a very efficient huntress. The important question of the relative contribution of experience and inheritance to this efficiency has hardly been explored, but it could be investigated by making careful observations on individually marked wasps and timing various activities such as hunting or depositing prey in the nest. Does an experienced wasp work faster than a novice?

Prey specificity seems to change slowly in evolution, and whole groups of wasps share similar tastes. Evans (1963) has suggested that in general the more ancient groups of wasps take those groups of prey that appear earliest in the fossil record. As well as offering scope for evolutionary speculation, prey selection can be a useful aid to rough identification because wasps are so consistent and so rarely make mistakes. If you see a wasp returning to its nest and can identify the group to which its prey belongs, you can go some way towards naming the wasp, and if you take nest type into account too you can make an even more accurate guess (see Guessing Guide, p. 28).

Egg numbers

We have seen that wasps show wide variation in the amount of time and energy they devote to preparing and provisioning the nest, and a correspondingly wide variation in the degree of protection of the young from damage by predators, pathogens and microclimatic extremes. The wandering wasps that make no nests must have more energy left to invest in eggs, and must suffer a high larval mortality, whereas a wasp like *Ammophila pubescens* devotes so much of her time and energy to each nest that she can lay few eggs but each has a high chance of survival. Iwata (1942) has shown a correlation between the number and size of a wasp's eggs and the type of nesting behaviour. In general his preliminary evidence indicated that wandering wasps and kleptoparasites lay numerous small eggs (averaging about 60 each, that is ten in each of the six ovarioles, chapter 4), whereas the hunting wasps with more advanced parental care lay fewer larger eggs (about two to five in each of the six ovarioles).

Nest-sharing and the path to sociality

If social wasps evolved from solitary species, it is among the solitary wasps that we should look for clues to the evolutionary path that led to sociality. If we are to recognise these clues we must identify the features that characterise the truly social insects. According to Wilson (1975) these are three: individuals co-operate in caring for the young; there is a worker caste that is sterile, or at least less fertile than the queen; and overlap of generations allows daughters to co-operate with their mothers in the work of the colony.

Eberhard (1978) suggested that the early stages in the evolution of sociality are represented by groups of females of the same generation, probably but not necessarily close relatives, sharing a nest. All or many of them are capable of laying eggs. She found records of two or more females sharing and provisioning a common nest in about 40 species of solitary wasps, and nearly all cases involved groups of reproductive females, which were often related. The largest or oldest female may reduce or suppress oviposition by less dominant females for a time, but a caste of sterile workers that do not lay eggs is found only in the truly social species.

Nest-sharing in solitary wasps is an essential prerequisite for primitive sociality. Eberhard's list of nest-sharers included several genera found in Britain: *Cerceris, Diodontus, Lindenius, Crossocerus, Stigmus, Spilomena, Ectemnius* and *Ammophila*. Nest-sharing is not easy to recognise, even in an intensive study of a colony of individually marked wasps. Two or more females using the same entrance hole may be sharing; but one may be usurping the other's nest, as *Cerceris arenaria* sometimes does, or stealing its provisions, as *Passaloecus corniger* does. Nest-sharing females cohabit over a period of time and tolerate one another's presence in the nest.

An opportunity to study a nest-sharing species is an opportunity to address some key questions posed by theoreticians but not yet answered by field-workers. It is a cornerstone of Eberhard's argument that nest-sharing is of mutual advantage to the females involved, perhaps because a shared nest is defended from parasites for more of the time than one used by a single female. This could be tested by comparing for shared and unshared nests the number of times parasitic wasps or flies enter the nest (by long-term observation) or the proportion of parasitised cells (by opening a nest: technique, p. 58).

Research

In this chapter we have tried to show how wasp behaviour patterns are worth studying at two levels: the broad strategy that matters to the wasp, and the fine detail that matters to the evolutionist. Evolutionary speculation serves to focus the attention of field-workers on significant aspects of behaviour, so that in due course the accumulation of relevant observations will enable the ideas to be critically re-examined and perhaps revised. With the benefit of knowledge gained from the earlier workers in this field, we can now see some of the gaps in the fabric so that today's wasp-watchers, with a clearer idea of what to look for, are less likely to overlook important details. Some of the questions of interest in a comprehensive behavioural study of a species are given in table 1 as a sort of checklist for field-workers; the reason why each question matters should be apparent from this chapter. Observations have been published on some of these questions for some of our species of wasps, but there are few species (notably *Ammophila pubescens* studied by Baerends in Holland and *A. sabulosa* studied by Field in Norfolk) for which all or even most of them have been answered. This is a limitless field of enquiry in which the careful amateur has an important contribution to make. Finally, we should say a word about ways of describing behaviour. When we talk about human behaviour we often describe it subjectively in terms of the supposed purpose of the action. We can do this because most human behaviour has a purpose which we can reasonably infer. But we cannot describe animal behaviour this way because we cannot impute to animals an intention or purpose. It is essential to use neutral, or objective, terms in the scientific description of animal behaviour, describing only, and exactly, what we see. Thus the primary account of behavioural observations must be scrupulously objective, even if subsequent more popular accounts derived from it are written in a more relaxed and subjective way to make them easier to read. Scrupulous care with the use of words is particularly important when describing the activities of wasps, in which trivial-seeming details can be valuable clues in tracing evolutionary patterns. Some of these details are difficult to see, and it is very important to make a clear distinction between certainties and guesses. Good photographs can help.

Table 1. A behavioural inventory for a species of solitary wasp: some questions

Nest
Does the wasp use a nest at all?
Does it prepare its own or use one made
  by another insect?
Is the nest prepared before or after
  hunting?
Does it nest in soil or in wood or stems?
If the nest is in soil, does excavation
  involve pushing, pulling, raking or
  carrying?
Is there more than one cell per nest?
If so, are the cells arranged in a row, or
  does the hole branch?
What are the internal partitions and the
  terminal closure made of?
Does the wasp block the hole
  temporarily when it leaves?
Is there evidence for scent-marking?

Prey
What is the prey?
What is the range of prey species used
  by an individual, or by all the wasps
  in one colony?
What is the sex ratio of the prey?
Do individual females specialise in
  particular prey species?
How many prey items are there per cell?
Do female-producing cells contain more
  prey items than male-producing
  cells?
Does the wasp hunt by ambush, or on
  foot, or in flight?
How does it paralyse the prey: by
  stinging, or by biting?
Is the prey transported to a nest?
Does the wasp carry it on foot or in
  flight?
How does the wasp grasp the prey
  during transport: with mandibles;
  legs only; or on the sting?
Does the wasp put the prey down at the
  nest entrance to readjust her grip or
  unblock the nest?
Does the wasp reclaim dropped prey?
Does a wasp revisit a cell and replenish
  the prey after that egg has hatched?

Egg-laying
Does the wasp lay her egg before or
  after bringing the first prey item to
  each cell?
Is the egg laid on the first, intermediate
  or last prey item in each cell?
Whereabouts on the prey's body is the
  wasp's egg laid?
How big is the egg in relation to the
  wasp?
How many eggs can one female lay?

Other adult behaviour
How do males find females?
Does mating take place at the nest site?
Do females mate once, occasionally or
  often?
Do males hold territories?
Do males help to protect nests from
  parasites?
Do adults feed on prey, or on flowers or
  on honeydew?
Where do male and female wasps spend
  the night?
Are nests grouped into colonies?
Do females ever share nests?
Do wasps steal each others' provisions
  or usurp each others' nests?

Nest associates
What other insects are associated with
  the nest colony?
Do other solitary wasps or bees nest
  nearby?
What other insects enter wasps' nest
  holes? Ruby-tailed wasps? Flies?
  Thieving, usurping or
  kleptoparasitic wasps?
Are there any kleptoparasites that lay
  eggs on the prey before the wasp
  carries it into her cell?
Are eggs or larvae of parasites or
  kleptoparasites to be found in the
  wasps' nests?
Do these larvae feed on the stored
  provisions or on the wasp's larva
  itself?
What proportion of wasps' cells are
  destroyed by parasites or
  kleptoparasites?
Is it possible to rear the parasites or
  kleptoparasites to the adult stage so
  that they can be identified?

3 Identification

It may be possible to make an approximate identification of a living wasp in the field, perhaps even without catching it. For this purpose use the Guessing Guide (p. 28). For wasps that have been killed or anaesthetised (technique, p. 57) use the Quick-Check Key to genera (p. 24) or the main keys which lead to species (pp. 26–56).

Some flies (Diptera), and even some moths, look like Hymenoptera. (In many cases this is protective mimicry.) So before starting on the keys, check with p. 1 that your insect is a hymenopteran.

Fig. 21. Glossary diagram

A wing vein is specified by the numbers designating the cells on either side; thus vein 3/4 separates cell 3 from cell 4.

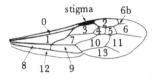

Cells 3, 4 and 5 are the submarginal cells.

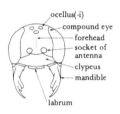

The main keys

These keys provide for the identification of all the genera and almost all the species in the groups covered by this book (see p. 64 and Alphabetical Checklist, p. 65). If you start with a member of a group that is not included, the keys will tell you. Note that in one of the included groups there are species in which the females are wingless (pls. 1 and 7).

Begin your examination by sexing your wasp. A female has 12 antennal and 6 abdominal segments. A male has 13 antennal segments (rarely 12, but the keys allow for this) and 7 abdominal. Then read both leads of the first couplet (i.e. '1' and ' – '), decide which corresponds better with your insect and follow the direction to the couplet indicated by the number on the right. Bracketed statements in the keys (other than instructions and special information) apply either to less reliable characters or to characters whose states are inconstant in the alternative lead and therefore not stated there. Beginners must start with Key I, which works differently: start at the top left-hand corner and follow the arrows, choosing one of the alternatives at each branch.

Dorsal means on or about the animal's back (as seen from above when it walks on a horizontal surface), and *ventral* means on or about the surface you would see from below an animal walking horizontally. The *length* of an insect means the combined length of head, thorax and abdomen, omitting mouthparts, antennae, ovipositor (egg-tube) and legs. *Pits* on the body surface are small pinprick-like impressions. The bracketed *dimensions in the keys* are lengths.

Most of the insects in the keys belong to the family Sphecidae (of the superfamily Apoidea); where they do not, we tell you. Groups not dealt with in this book appear in square brackets. Numbers in quotation marks in the keys are cross-references to other couplets or leads.

Always check the probability of a correct identification against the illustrations (pls. 1–8) and against the geographical distribution in the Alphabetical Checklist (p. 65). If you intend to publish work on a wasp, it is wise to make sure of its identification by consulting an expert or by checking with the more detailed descriptions and illustrations in Richards (1980), and perhaps also in de Beaumont (1964) or Lomholdt (1975).

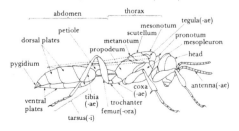

Quick-Check Key*

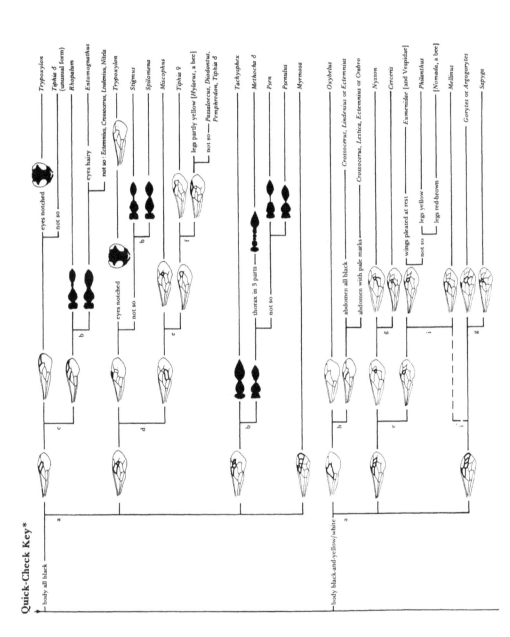

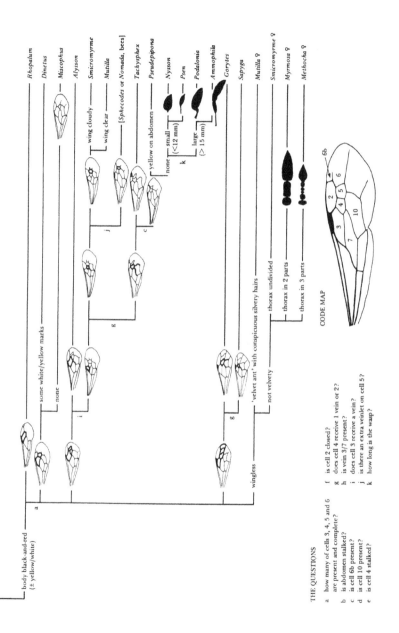

THE QUESTIONS

a how many of cells 3, 4, 5 and 6
   are present and complete?
b is abdomen stalked?
c is cell 6b present?
d is cell 10 present?
e is cell 4 stalked?

f is cell 2 closed?
g does cell 4 receive 1 vein or 2?
h is vein 3/7 present?
i does cell 3 receive a vein?
j is there an extra veinlet on cell 5?
k how long is the wasp?

CODE MAP

* This includes only the wasps covered by this book (and a few other things, in square brackets, that might be confused with them). It is partly based on Shuckard (1837).

Key I (please see p. 23 before you start)

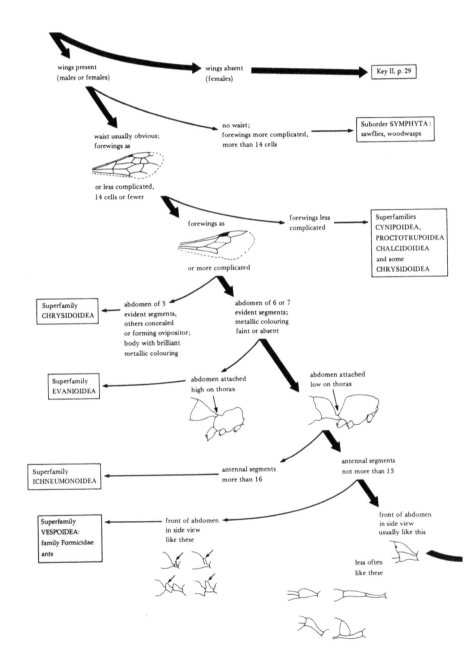

wings present
(males or females)

wings absent
(females)

Key II, p. 29

waist usually obvious;
forewings as

no waist;
forewings more complicated,
more than 14 cells

Suborder SYMPHYTA:
sawflies, woodwasps

or less complicated,
14 cells or fewer

forewings as

forewings less
complicated

Superfamilies
CYNIPOIDEA,
PROCTOTRUPOIDEA
CHALCIDOIDEA
and some
CHRYSIDOIDEA

or more complicated

Superfamily
CHRYSIDOIDEA

abdomen of 3
evident segments,
others concealed
or forming ovipositor;
body with brilliant
metallic colouring

abdomen of 6 or 7
evident segments;
metallic colouring
faint or absent

Superfamily
EVANIOIDEA

abdomen attached
high on thorax

abdomen attached
low on thorax

Superfamily
ICHNEUMONOIDEA

antennal segments
more than 16

antennal segments
not more than 13

Superfamily
VESPOIDEA:
family Formicidae:
ants

front of abdomen
in side view
like these

front of abdomen
in side view
usually like this

less often
like these

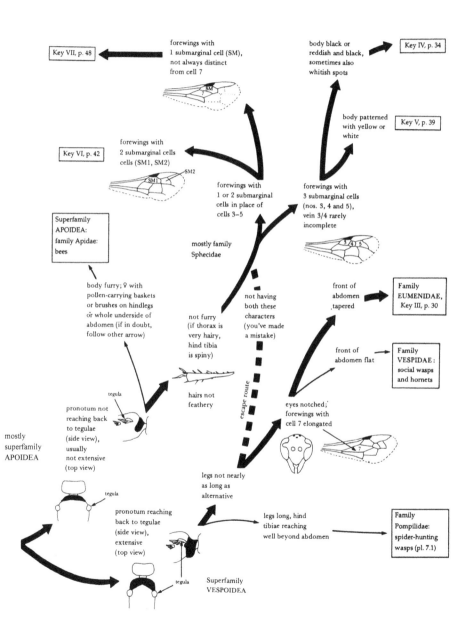

Key VII, p. 48

forewings with
1 submarginal cell (SM),
not always distinct
from cell 7

body black or
reddish and black,
sometimes also
whitish spots

Key IV, p. 34

body patterned
with yellow or
white

Key V, p. 39

Key VI, p. 42

forewings with
2 submarginal cells
cells (SM1, SM2)

SM2

SM1

Superfamily
APOIDEA:
family Apidae:
bees

forewings with
1 or 2 submarginal
cells in place of
cells 3–5

mostly family
Sphecidae

forewings with
3 submarginal cells
(nos. 3, 4 and 5),
vein 3/4 rarely
incomplete

3/4 5

body furry; ♀ with
pollen-carrying baskets
or brushes on hindlegs
or whole underside of
abdomen (if in doubt,
follow other arrow)

not furry
(if thorax is
very hairy,
hind tibia
is spiny)

not having
both these
characters
(you've made
a mistake)

front of
abdomen
tapered

Family
EUMENIDAE,
Key III, p. 30

front of
abdomen flat

Family
VESPIDAE:
social wasps
and hornets

tegula

hairs not
feathery

escape route

eyes notched;
forewings with
cell 7 elongated

pronotum not
reaching back
to tegulae
(side view),
usually
not extensive
(top view)

mostly
superfamily
APOIDEA

legs not nearly
as long as
alternative

tegula

pronotum reaching
back to tegulae
(side view),
extensive
(top view)

legs long, hind
tibiae reaching
well beyond abdomen

Family
Pompilidae:
spider-hunting
wasps (pl. 7.1)

tegula

Superfamily
VESPOIDEA

Guessing Guide

    This Guide is meant to help name wasps seen carrying prey to their nests. To use it, see whether the nest is in soil, or in wood (or a stem or twig). See whether the wasp's body (thorax and abdomen) is all black; or black marked with yellow or white only; or black marked with red (with or without (±) yellow or white). See what prey the wasp carries. (It may be necessary to catch or frighten her to confiscate her prey.) From this information the Guide will give a short list of possible genera, some of which may then be eliminated on grounds of size (given in the main keys, page references in the Alphabetical Checklist, p. 64) or appearance (pls. 1 – 8). *Never* trust an identification made with this Guide; as soon as possible, take a specimen for more critical identification using the main key.

Prey type	Nests in wood, stems, twigs			Nests in soil, masonry		
	Body black + red (± yellow/white)	Body black + yellow/white	Body all black	Body black + red (± yellow/white)	Body black + yellow/white	Body all black
Spiders			*Trypoxylon*	*Miscophus*		*Miscophus*
Flies	*Rhopalum*	*Crossocerus* *Ectemnius*	*Crossocerus* *Rhopalum*		*Lindenius* *Mellinus* *Oxybelus* *Crabro* *Crossocerus*	*Lindenius* *Crossocerus*
Other	*Rhopalum*	*Crossocerus* *Lestica*	*Spilomena* *Rhopalum* *Nitela*		*Lindenius* *Crossocerus*	*Lindenius*
Homoptera	*Rhopalum*		*Crossocerus* *Psen* *Psenulus* *Passaloecus* *Pemphredon* *Stigmus* *Nitela*	*Gorytes* *Psen* *Alysson*	*Gorytes*	*Diodontus*
Adult beetles					*Entomognathus* *Cerceris*	*Entomognathus*
Heteroptera				*Astata* *Dinetus*		*Lindenius*
Grasshoppers or cockroaches				*Tachysphex*		*Tachysphex*
Bees					*Philanthus* *Cerceris*	
Beetle larvae		*Ancistrocerus* *Symmorphus* *Gymnomerus*		*Methocha*	*Ancistrocerus* *Odynerus*	*Tiphia*
Caterpillars		*Ancistrocerus* *Symmorphus* *Eumenes*		*Pseudepipona* *Podalonia* *Ammophila*	*Ancistrocerus* *Euodynerus* *Pseudepipona* *Eumenes*	
Parasites or kleptoparasites	*Sapyga*	*Sapyga*		*Nysson* *Myrmosa* *Mutilla* *Smicromyrme*	*Nysson*	

II Wingless Hymenoptera (female only)

1   Front of abdomen as in II.1, II.2 or II.3
[superfamily VESPOIDEA: family Formicidae (ants)]
–   Front of abdomen otherwise   2

II.1

2   Abdomen flattened from side to side (insect not more than 4 mm)
[superfamily CYNIPOIDEA (gall-wasps and allies)]
–   Abdomen not flattened from side to side (insect not less than 3.5 mm)   3

3   Antennal segments 16 or more
[superfamily ICHNEUMONOIDEA]
II.2
–   Antennal segments 12 or 13   4

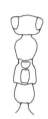

4   Thorax with no clear divisions but a faint one marking the propodeum; dorsal plates 1–3 or 2–3 patterned with patches of white silky hairs   5
–   Thorax with distinct divisions (II.4, II.5); abdomen with no pattern formed by hairs, or only complete bands   6

II.3

5   Length 7.5–15 mm; 1st dorsal plate more than half as wide as 2nd (pl. 1.1)
(superfamily VESPOIDEA: family Mutillidae)
Mutilla europaea
–   Length 3–6 mm; 1st dorsal plate less than half as wide as 2nd (pl. 7.6)   (family Mutillidae) *Smicromyrme rufipes*

II.4

6   Thorax as II.4; body highly polished, very sparsely pitted; (3.5–8.5 mm) (pl. 7.7)
(superfamily VESPOIDEA: family Tiphiidae)
Methocha ichneumonides
–   Thorax as II.5; body duller, closely pitted on head, thorax and 1st dorsal plate; (3.5–6 mm) (pl. 1.2)
(family Tiphiidae) *Myrmosa atra*

II.5

III Species of the family Eumenidae: potter wasps and mason wasps (superfamily VESPOIDEA)

III.1   III.2

1   Front of abdomen as in III.1; (9–15 mm) (pl. 8.1)
                                             Eumenes coarctatus
–   Front of abdomen as in III.2   2

III.3   III.4

2   Tegulae rounded behind (III.3); ♂ tip of antenna loosely
    curled (III.4)   3
–   Tegulae pointed behind (III.5); ♂ tip of antenna either
    tightly hooked (last 2 segments very small) or normal
    (III.6)   7

III.5

3   ♂ middle coxa with a large tooth (III.7); ♂ with lobes
    behind mandibles (face view) (III.8); ♀ with pale spot on
    metanotum   4
–   ♂ middle coxa without tooth; ♀ without lobes behind
    mandibles; ♀ without mark on metanotum   5

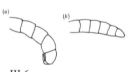

III.6

4   Markings creamy white; ♂ lobes behind mandibles
    black; ♀ without spots on propodeum; (♂ about 11 mm;
    very rare)   *Odynerus simillimus*
–   Markings yellow; ♂ lobes behind mandibles brown; ♀
    with large spots on propodeum; (apparently extinct;
    9–12 mm)   *Odynerus reniformis*

III.7

5   Front of thorax as in III.9; ♂ middle femora not as in
    III.10; (markings pale yellow); (8–11 mm)
                                          Gymnomerus laevipes
–   Front of thorax as in III.11, with furrow about parallel to
    sides of thorax (less distinct in ♂ than ♀); ♀ middle
    femora as in III.10 (seen from above and behind);
    (markings creamy white or yellow)   6

III.8

6   2nd–4th ventral plates with pairs of spots (usually
    whitish) decreasing regularly in size towards rear; ♀
    with 4 dorsal bands; ♀ with lower front of antennae
    whitish; (8–10 mm; rare) (pl. 3.6)
                                     Odynerus melanocephalus
–   2nd ventral plate with an interrupted band (yellow) or
    unmarked, 3rd with a pair of small spots or unmarked,
    4th unmarked; ♀ with 5 dorsal bands; ♀ with antennae
    black; (8–12 mm)   *Odynerus spinipes*

III.9

III.10

III.11

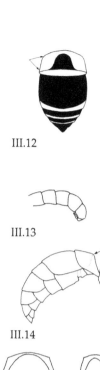

III.12

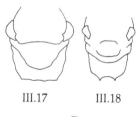

III.13

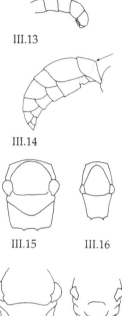

III.14

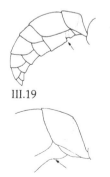

III.15   III.16

III.17   III.18

III.19

III.20

7   Front of abdomen with an orange–brown area (III.12);
    (other body markings creamy white; ♂ antennae as in
    III.13); (9–11 mm; very rare)   *Pseudepipona herrichii*
–   Front of abdomen without orange-brown area   8

8   1st dorsal plate with a ridge or angulation across it
    (III.14) (this can be hard to see: tilt the specimen about)
                                                            10
–   1st dorsal plate with no such ridge or angulation   9

9   Thorax as in III.15; markings yellow; (9–11 mm; very rare)
                                          Euodynerus quadrifasciatus
–   Thorax as in III.16; markings creamy white; (♂ antennae
    as in III.13); (6–8 mm)   *Microdynerus exilis*

10 Thorax angled behind as in III.17 (view from above the
    metanotum); 1st dorsal plate without a lengthwise
    groove on the rear part; ♂ antennae as in III.13
                                          Ancistrocerus 11
–   Thorax behind as in III.18; 1st dorsal plate with a
    lengthwise groove on the rear part; ♂ antennae not as in
    III.13   *Symmorphus* 20

11 2nd ventral plate as in III.19; (1st dorsal plate with the
    yellow band broad, with a rectangular notch; ♂ with 6
    bands, ♀ with 5; ♀ clypeus with 2 or 4 yellow spots);
    (9–13 mm) (pl. 3.5)   *Ancistrocerus nigricornis*
–   2nd ventral plate not definitely stepped
    (III.20, III.24 or III.25)   12

12 2nd ventral plate from below as in III.21 (if hairs
    interfere with view alter the light), its side view slightly
    concave (III.20) or straight   13
–   2nd ventral plate from below as in III.22, its side view
    gently to abruptly convex (III.25 or III.24)   15

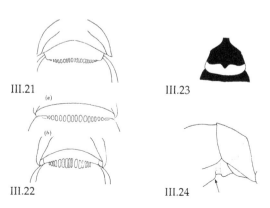

III.21
(a)

(b)

III.22

III.23

III.24

13 *The characters given here for* A. parietum *and* A. gazella
    *and additional ones in other works are inconstant; in
    doubtful cases more detailed keys may permit a decision*
    Ridge across 1st dorsal plate with a forwardly directed,
    broadly V-shaped central notch (view from front);
    yellow on 1st dorsal plate gradually widened at sides; ♂
    front corners of pronotum not produced into sharp teeth;
    (♂ 1st–6th dorsal plates with a yellow band,
    ♀ 1st–5th with band, 6th usually with a spot, otherwise
    black); (7.5–12 mm)   *Ancistrocerus parietum*
–   Ridge across 1st dorsal plate without a forwardly
    directed notch but sometimes knotted at the middle;
    yellow on 1st dorsal plate with a rectangular
    enlargement at the sides; ♂ front corners of pronotum
    produced into sharp teeth, sometimes projecting
    diagonally   14

III.25

14 Forewings distinctly clouded in cells 1 and 2; ♂ and ♀
    1st–4th or 5th dorsal plates with a yellow band, ♂ 6th
    sometimes with a spot; (7–12 mm)   *Ancistrocerus gazella*
–   Forewings only faintly clouded in cells 1 and 2 (but if
    wings are folded they may seem dark); ♂ 1st–6th dorsal
    plates with a yellow band, ♀ 1st–5th with band, 6th with
    a spot; (7.5–10 mm, rare)   *Ancistrocerus quadratus*

III.26

15 Sides of propodeum smooth and shining, rear surface
    partly shining; (yellow bands on dorsal plates
    numbering 4–6 in ♂, 4–5 in ♀; ♀ 1st band as in III.23 or
    straight); (11–16 mm, rare)   *Ancistrocerus antilope*
–   Sides of propodeum rough above, finely ridged
    horizontally below, dull, rear surface with strong to
    weak diagonal striations   16

III.27

16 Light bands on dorsal plates numbering 5 or 6   17
–   Light bands on dorsal plates numbering 3 or 4, a portion
    of a 5th sometimes present   19

17 A male with yellow on face beside eye
                              Ancistrocerus nigricornis (see '11')
–   A female; or a male without yellow marks on face beside
    eye   18

18 Usually no black on tibiae; 2nd ventral plate abruptly
    convex (III.24); ♂ clypeus notched as in III.26; ♀ clypeus
    black; (body markings yellow or white); (9–14 mm)
    (pl. 8.2)   *Ancistrocerus oviventris*
–   Some or all tibiae edged with black; 2nd ventral plate
    gently convex (III.25); ♂ clypeus notched as in III.27; ♀
    clypeus with yellow spots; (body markings yellow);
    (10–13.5 mm)   *Ancistrocerus parietinus*

PLATE 1

1
Mutilla europaea ♀

2
Myrmosa atra ♀

3
Sapyga quinquepunctata ♀

4
Astata pinguis ♂

5
Sphecodes monilicornis ♀
(bee)

6
Dinetus pictus ♀

7
Miscophus concolor ♀

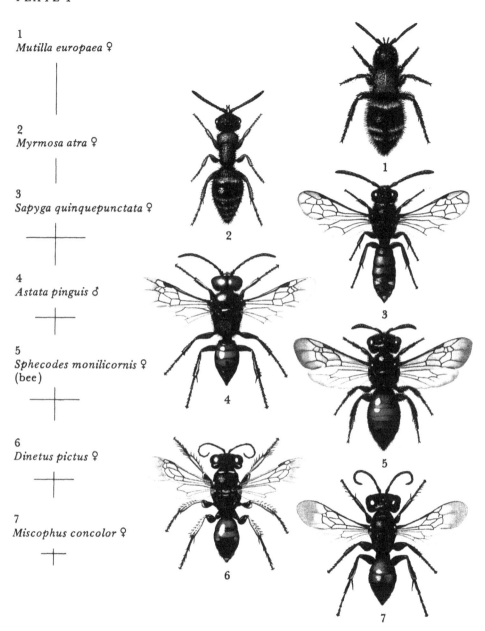

Scale lines show actual
size of insect

PLATE 2

1
Podalonia hirsuta ♀

2
Psen lutarius ♂

3
Nysson dimidiatus ♀

4
Nysson spinosus ♀

5
Gorytes tumidus ♀

6
Gorytes quadrifasciàtus ♀

7
Argogorytes mystaceus ♀

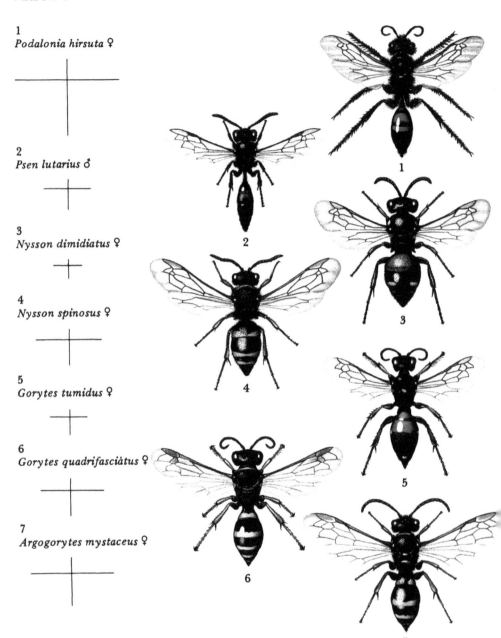

PLATE 3

1
Mellinus arvensis ♀

2
Cerceris rybyensis ♀

3
Nomada marshamella ♀
(bee)

4
Symmorphus bifasciatus ♀

5
Ancistrocerus nigricornis ♀

6
Odynerus melanocephalus
♀

7
Oxybelus uniglumis ♀

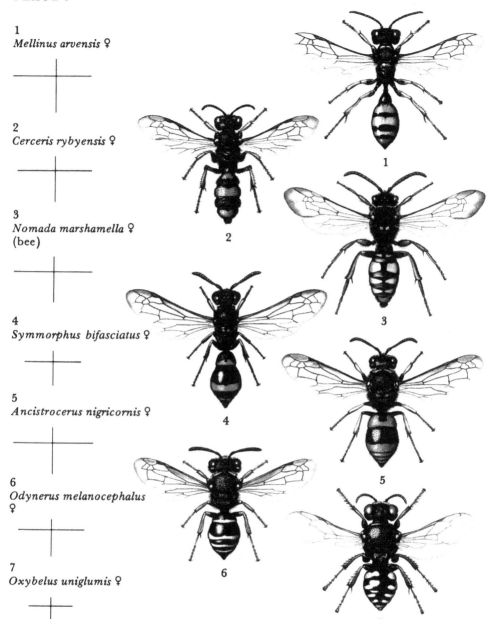

PLATE 4

1
Crabro cribrarius ♀

2
Crabro cribrarius ♂

3
Ectemnius lapidarius ♀

4
Ectemnius continuus ♀

5
*Crossocerus
quadrimaculatus*
pale form ♀

6
*Crossocerus
quadrimaculatus*
dark form ♀

7
Rhopalum clavipes ♀

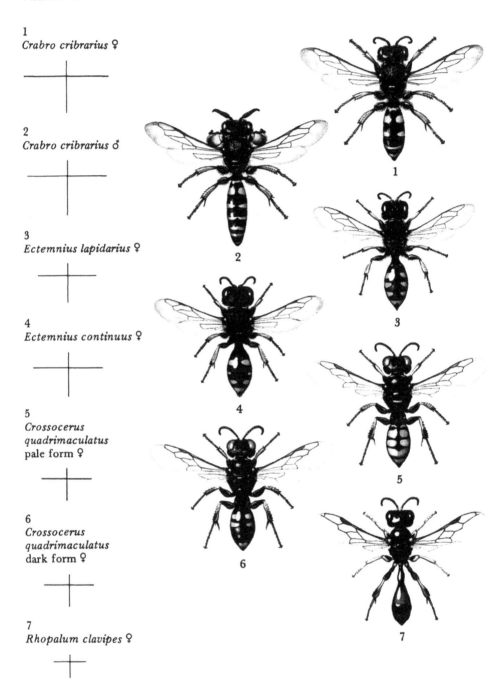

PLATE 5

1
Myrmosa atra ♂

2
Psenulus pallipes ♀

3
Hylaeus hyalinatus ♀
(bee)

4
Diodontus minutus ♀

5
Spilomena troglodytes ♀

6
Pemphredon lugubris ♀

7
Tiphia femorata ♀

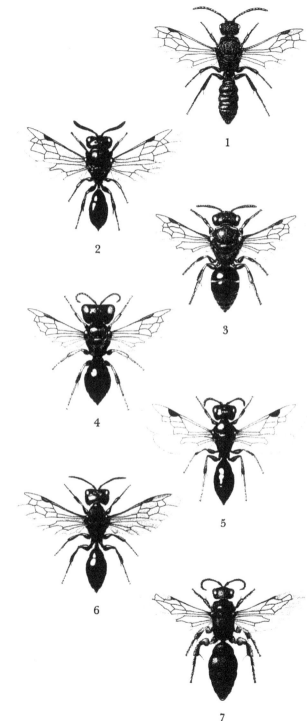

1

2

3

4

5

6

7

PLATE 6

1
Passaloecus corniger ♀

2
Trypoxylon figulus ♂

3
Entomognathus brevis ♀

4
Lindenius albilabris ♀

5
Crossocerus megacephalus ♀

6
Crossocerus elongatulus ♀

7
Crossocerus wesmaeli ♀

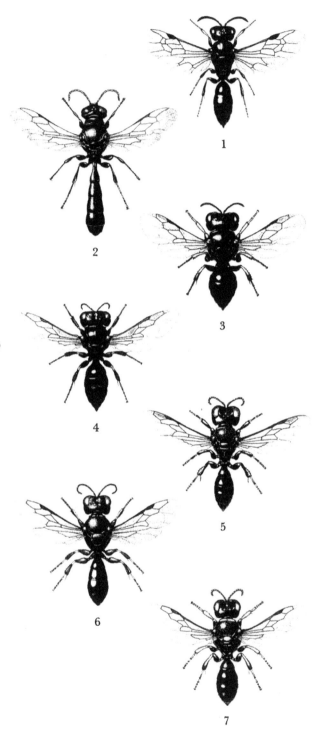

PLATE 7

1
Priocnemis exaltatus ♂
(spider-hunting wasp)

2
Alysson lunicornis ♀

3
Tachysphex pompiliformis ♀

4
Ammophila pubescens ♀

5
Astata boops ♀

6
Smicromyrme rufipes ♀

7
Methocha ichneumonides ♀

8
Sapyga clavicornis ♀

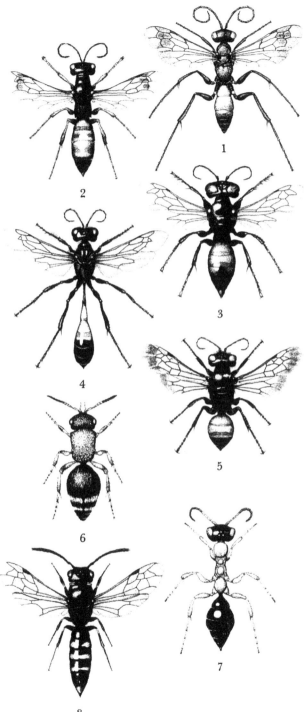

PLATE 8

1
Eumenes coarctatus ♀

2
Ancistrocerus oviventris ♀

3
Cerceris arenaria ♀

4
Gorytes bicinctus ♀

5
Philanthus triangulum ♀

6
Lestica clypeata ♂

7
Crossocerus vagabundus ♀

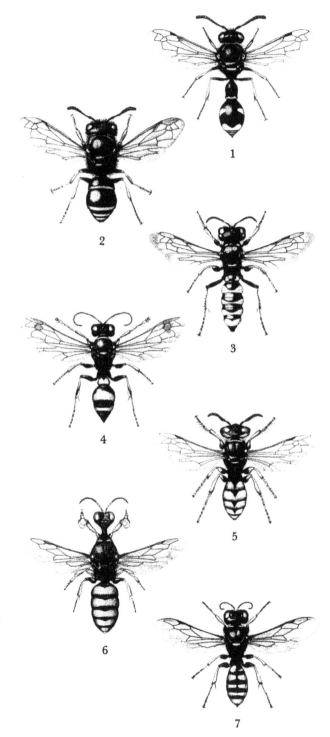

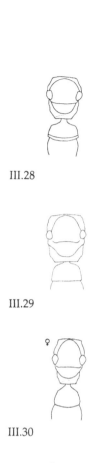

III.28

III.29

III.30

III.31

III.32

19 ♂ usually with no black on hind tibiae; ♀ legs largely
    yellowish-brown with black markings; stouter,
    especially ♂ thorax (III.28) and ♀ 1st abdominal segment
    (III.29); ♂ front corners of pronotum blunt (III.28);
    (markings yellow or white); (8.5–12 mm)
    Ancistrocerus scoticus

–   ♂ usually with black mark on hind tibiae; ♀ legs largely
    yellow, markings black and/or brown; slenderer, thorax
    and 1st abdominal segment as in III.30 and III.31; ♂ front
    corners of pronotum acute (III.30); (9–12 mm)
    Ancistrocerus trifasciatus

20 Head and thorax, seen from side, clothed with long,
    more or less curved, pale brown hairs; pronotal yellow
    patches as in III.32; (♂ 1st–5th or 6th dorsal plates with a
    yellow band, ♀ 1st–4th or 5th with band); (8–15 mm)
    Symmorphus crassicornis

–   Head and thorax, seen from side, with shorter, straight,
    black or whitish hairs; pronotal yellow patches as in
    III.33 or III.34, or absent   21

21 Pronotum not flanged along front edge or only so at
    sides; ♂ 1st antennal segment yellow on front; 3rd and
    4th dorsal plates with a yellow band (1st–5th or 6th
    banded in ♂, 1st–4th or 5th in ♀); (yellow on pronotum
    as in III.33); (7–12 mm)   *Symmorphus gracilis*

–   Pronotum flanged all across front edge (III.34); ♂ 1st
    antennal segment black; 3rd dorsal plate with at most 2
    yellow spots, 4th with or without a band   22

22 Mesonotum and mesopleuron freely or closely pitted; ♀
    scutellum 2-spotted; ♂ clypeus with at most a yellow
    patch at the top; (♀ yellow on pronotum as in III.34, not
    filling the corners; ♂ with none); (6.5–9.5 mm) (pl. 3.4)
    Symmorphus bifasciatus

–   Mesonotum and mesopleuron sparsely pitted; ♀
    scutellum black; ♂ clypeus normally yellow with a
    narrow black edge; (pronotum black); (6.5–9 mm; rare)
    Symmorphus connexus

III.33

III.34

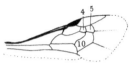

IV **Solitary wasps with three submarginal cells and body black or with red markings (superfamilies VESPOIDEA and APOIDEA)**

IV.1

1   Body black   2
–   Body with red markings   13

2   Forewings with cell 10 in contact with cell 5 but hardly with cell 4 (IV.1)
                    [superfamily APOIDEA: family Apidae (bees), *Halictus* and *Lasioglossum*]
IV.2
–   Forewings with cell 10 in contact with cell 4, more so than with cell 5 (IV.2 and IV.3)   3

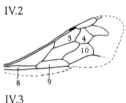

3   Forewings as in IV.2, most of vein 3/4 parallel to front of wing; body surface heavily pitted; (male, ♀ being wingless: see key II, '6– '; 5.5–11 mm) (pl. 5.1)
                    (superfamily VESPOIDEA: family Tiphiidae)
IV.3   *Myrmosa atra*
–   Forewings with vein 3/4 as in IV.3, at an angle of 45° or more to front of wing; body surface not heavily pitted   4

4   Front of abdomen rounded when seen from above; forewings with cell 8 not longer than 9 (IV.3); (4–7 mm)
IV.4   *Tachysphex unicolor*
    (*T. pompiliformis* and *T. obscuripennis* have red on abdomen: couplet '32')

–   Front of abdomen in the form of a narrow stalk (petiole); forewings with cell 8 much longer than cell 9   5

IV.5

5   Furrowed area on propodeum shaped as in IV.4; forewings with vein 10/11 running into cell 4 (IV.6)
                                        Psen subgenus *Mimumesa* 7
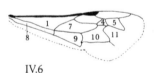
    (subgenus *Mimesa* has red on abdomen: couplet '20')
–   Furrowed area on propodeum as in IV.5; vein 10/11 running into cell 5 (IV.7) or joining vein 4/5   6

IV.6

6   Petiole of abdomen longer than swollen part of first segment; (9–13 mm; not found since end of 19th century)
                                        Psen (subgenus *Psen*) *ater*
–   Petiole of abdomen not longer than swollen part of 1st segment; 5–8 mm long   *Psenulus* 11

IV.7

7   Front of thorax beneath as in IV.8; ♀ head with an area just outside the lateral ocelli where the pits are rather sparse; ♂ tarsi yellow   8
–   Front of thorax beneath as in IV.9; ♀ head with no area of sparser pitting near the lateral ocelli; ♂ tarsi dark brown or black   9

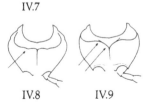

IV.8   IV.9

8   Head in front of ocelli with flat areas between the pits; ♂
petiole much longer than 1st dorsal plate (apparently
1¼–1½ times in a straight line along the top in side
view), ♀ petiole also relatively long; (6–8 mm)
Psen littoralis

–   Head in front of ocelli with the pits close, touching each
other; ♂ petiole not much longer than 1st dorsal plate
(observed as in lead '8'), ♀ petiole also relatively short;
(6–9 mm, rare)   *Psen unicolor*

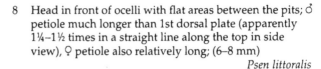

11   10   9   8

IV.10

9   Petiole in side view clearly longer than 1st dorsal plate;
♂ antennae with short bumps on the rear of segments
9–11 and a long but faint one on segment 8 (IV.10);
(7–11 mm)   *Psen atratinus*

–   Petiole in side view scarcely longer than 1st dorsal plate;
♂ antennae not like this   10

IV.11

10   Distance from lateral ocellus to eye about equal to
distance between centres of lateral ocelli; ♀ last dorsal
plate as in IV.11, shining; ♂ antennae with rather distinct
bumps on the rear of segments 4–10 or 4–11; (6–9 mm)
Psen dahlbomi

–   Distance from lateral ocellus to eye greater than distance
between centres of lateral ocelli; ♀ last dorsal plate as in
IV.12, dull; ♂ antennae with a ridge running the whole
length of the rear of segments 4–11, but faint and
difficult to see; (6–9 mm)   *Psen spooneri*

IV.12

11   Forehead distinctly pitted and/or furrowed   12

–   Forehead with only very faint pits, smooth and shining;
(♀ 4th and 5th ventral plates without a fringe of large
hairs); (6–8 mm)   *Psenulus concolor*

IV.13

12   ♀ middle tibia with no ridge and crest on outside; ♀ 4th
and 5th ventral plates with rear fringe of large hairs; ♂
antennal segments from above as in IV.13; (5–7 mm)
(pl. 5.2)   *Psenulus pallipes*

–   ♀ middle tibia with an oblique ridge and a blunt crest on
outer surface (IV.14); ♀ 4th and 5th ventral plates
without a fringe of large hairs; ♂ antennal segments 5–11
as in IV.15; (6–7 mm)   *Psenulus schencki*

IV.14

13   Thorax partly red-brown; rest of insect without pale
markings in the cuticle but with a pattern on the
abdomen formed by dense pale hairs; (insect very hairy)
14

–   Thorax usually black; if with brown markings there are
strong coloured markings in the cuticle of the abdomen
15

IV.15

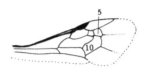

IV.16

IV.17

IV.18

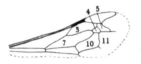

IV.19

IV.20

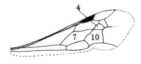

IV.21

IV.22

14 Length 9–14 mm; (male, ♀ being wingless: see key II, '5')
    (superfamily VESPOIDEA: family Mutillidae)
                                            Mutilla europaea
  – Length 4.5–7 mm; (male, ♀ being wingless: see key II, '5– ')
    (family Mutillidae) *Smicromyrme rufipes*

15 Forewings with cell 10 in extensive contact with cell 5
    (IV.16)   16
  – Forewings with cell 10 not in contact with cell 5 (IV.l9)
                                                        17

16 No white spots on body (except labrum); thorax as in
    IV.17   [superfamily APOIDEA: family Apidae (bees),
      Nomada, Halictus, Lasioglossum and *Sphecodes* (pl. 1.5)]
  – White spots on head, thorax and abdomen; front of
    thorax as in IV.18; (9–13 mm) (pl. 1.3)
                    (superfamily VESPOIDEA: family Sapygidae)
                                    Sapyga quinquepunctata
    (Another species of *Sapyga* is at key V, '20')

17 Forewings with cells 3 and 5 in contact at the front
    (IV.19); (propodeum with a spine at the top on either
    side (IV.20))   18
  – Forewings with cells 3 and 5 not in contact   19

18 Forewings with vein 7/10 as well as 10/11 joining cell 4
    (IV.19); spines of propodeum large; ♂ antennae normal;
    (red band on abdomen; white spots on thorax and
    abdomen); (4–6 mm) (pl. 2.3)   *Nysson dimidiatus*
    (Other species of *Nysson* are at key V, '10– ')
  – Forewings with vein 7/10 not joining cell 4 (IV.21);
    spines of propodeum small; ♂ antennae as in IV.22;
    (5.5–8 mm, rare) (pl. 7.2)   *Alysson lunicornis*

19 Front of abdomen in the form of a stalk (petiole)   20
  – Front of abdomen rounded or shortly tapered   30

20 Vein 7/10 meeting or nearly meeting vein 3/4; stigma
    about ¾ as long as cell 3; insect not more than 10 mm
                            Psen subgenus *Mimesa* 21
    (subgenus *Mimumesa* has no red on abdomen: couplet '5')
  – Vein 7/10 not nearly meeting vein 3/4; stigma about
    ⅕–¼ as long as cell 3; insect 13–24 mm   27

21 Antennae with some segments near the tip 1½ times as
    long as broad (longer than this in ♂), segment 3 in ♀ 4
    times as long as broad; petiole conspicuously longer
    than 1st dorsal plate; (♀ 1st dorsal plate with black
    patch); (7–10 mm)   *Psen bruxellensis*
–   Antennae with segments near the tip at most a little
    longer than broad, segment 3 in ♀ not more than 3 times
    as long as broad; petiole at most a little longer than 1st
    dorsal plate   22

22 Males (antennae with 13 segments; tip of abdomen with
    a deceptive sting-like spine)   23
–   Females (antennae with 12 segments)   25

23 Antennae seen from above without bulges on rear of
    segments (but hind profile may appear stepped at tips of
    segments 3–5); (petiole with a low ridge along the top;
    side of mesothorax as in ♀: lead '26– '); (6–10 mm)
                                                Psen bicolor
–   Antennae seen from above with bulges on rear of middle
    segments (representing lengthwise ridges)   24

24 Antennae with bulges on segments 3–7 or 3–8; petiole
    with a ridge along the top; tibiae usually with brown
    markings; (side of mesothorax as in ♀: lead '25');
    (6–9 mm)   *Psen equestris*
–   Antennae with bulges on segments 4–9 or 4–10; petiole
    with no lengthwise ridge, sometimes a slight hollow;
    tibiae black, almost without brown markings; (side of
    mesothorax as in ♀: lead '26'); (7–9 mm) (pl. 2.2)
                                            Psen lutarius
25 1st dorsal plate with a black patch at front; side of
    mesothorax (below wings) sparsely and faintly pitted
    and finely ridged, the two effects usually not visible at
    the same angle; (petiole with a ridge along the top; 3rd
    dorsal plate with red on it); (5–8 mm)   *Psen equestris*
–   1st dorsal plate normally without black patch; sides of
    mesothorax distinctly or heavily pitted and ridged, the
    two effects usually visible at the same angle   26

26 3rd dorsal plate usually without red on it; sides of
    mesothorax with distinct but well-spaced pitting and
    ridging, the pits more easily seen than the ridges; tibiae
    black except for a brown mark near tip of hind ones; (7–9
    mm) (pl. 2.2)   *Psen lutarius*
–   3rd dorsal plate with red on it; sides of mesothorax
    strongly pitted and closely and strongly ridged, the
    ridges more easily seen than the pits; tibiae all with
    some pale markings; (6–9 mm)   *Psen bicolor*

IV.23

27 Petiole abruptly set off from the swollen part of the
abdomen (males must be viewed from the side) (IV.23)
Podalonia 28
–   Petiole enlarging gradually into the swollen part of the
abdomen (IV.24)   *Ammophila* 29

28 Top of propodeum with the wrinkles forming a fine
network; (14–23 mm) (pl. 2.1)   *Podalonia hirsuta*
–   Top of propodeum with parallel wrinkles running
diagonally backwards from the midline; (13–20 mm)
Podalonia affinis

IV.24

29 Forewings with cell 4 in contact with cell 6 (IV.25); rear
part of abdomen without metallic blue sheen;
(13–19 mm)   *Ammophila pubescens*
–   Forewings with cell 4 not in contact with cell 6 (IV.26);
rear part of abdomen with a faint metallic blue sheen;
(14–24 mm)   *Ammophila sabulosa*

IV.25

30 White spots present on thorax and abdomen; forewings
with no trace of cell 6b; (7–8 mm) (pl. 2.5)
Gorytes tumidus
(Other species of *Gorytes* are at key V, '16')
–   Thorax and abdomen without white spots; cell 6b
evident (IV.27 and IV.28)   31

IV.26

31 Forewings with stigma indistinct, scarcely wider than
adjoining veins (IV.27)   *Tachysphex* 32
–   Forewings with a distinct stigma, much wider than
adjoining veins (IV.28); (hindwings about as wide as
forewings)   *Astata* 33

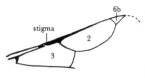

IV.27

32 4th segment of hind tarsi not broader than long; fore-
tibiae blackish beneath; ♂ facial hairiness silvery;
(5–7 mm) (pl. 7.3)   *Tachysphex pompiliformis*
(*T. unicolor* has no red on abdomen: couplet '4')
–   4th segment of hind tarsi broader than long; fore-tibiae
reddish to yellowish beneath; ♂ facial hairiness golden;
(6–10 mm; known from a single ♂ in 1882)
Tachysphex obscuripennis

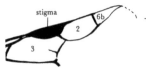

IV.28

33 Length 6–8 mm; ♀ pygidium (last dorsal plate) strongly
shining, not bordered by spines; ♂ with a large white
spot on face (pl. 1.4)   *Astata pinguis*
–   Length 9–13 mm; ♀ pygidium dull, bordered with spines
(IV.29); ♂ face black (pl. 7.5)   *Astata boops*

IV.29

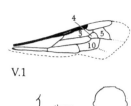

V.1

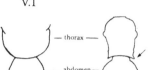

V Solitary wasps with three submarginal cells and yellow or white body pattern (superfamilies VESPOIDEA and APOIDEA)

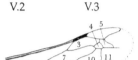

V.2   V.3

V.4

1   Forewings with cells 3 and 5 in contact at the front (V.1)   2
–   Forewings with cells 3 and 5 not in contact (V.18)   13

2   Forewings with cell 10 in contact with cells 4 and 5 (V.1); rear of propodeum rounded (V.2)   *Cerceris* 3
–   Forewings with cell 10 in contact with cell 4 but not with cell 5 (V.4); propodeum with a spine at the top on either side (V.3)   10

3   2nd dorsal plate with the yellow band at the front   4
–   2nd dorsal plate with the yellow band at the rear   5

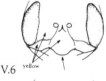

V.5

4   4th dorsal plate with less yellow on it than 5th; (♂ 6–10 mm, ♀ 8–12 mm) (pl. 3.2)   *Cerceris rybyensis*
–   4th and 5th dorsal plates with similar yellow bands; (6–10 mm; known from a single ♀ in 1861)
                                                          Cerceris sabulosa

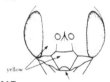

V.6 yellow

5   Female   6
–   Male   7

6   Clypeus (its middle part, in fact) standing out in front of the face as in V.5; (10–13 mm)   *Cerceris ruficornis*
–   Lower face as in:
    V.6; (8–12 mm) (pl. 8.3)   *Cerceris arenaria*
    V.7; (rare; 7–10 mm)   *Cerceris quinquefasciata*
    V.8; (very rare; 8–10 mm)   *Cerceris quadricincta*

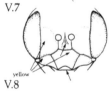

yellow

V.7

7   6th ventral plate with a dense brush-like tuft of bristles at either side, visible from above or from sides (insect very hairy in this area but other hairs not matted)   8
–   6th ventral plate with no brush-like tuft lying alongside 7th segment   9

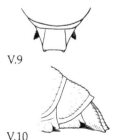

yellow

V.8

8   Hair-tufts on 6th ventral plate reddish-brown, slightly divergent, stronger, more easily seen from above (V.9); lower edge of clypeus with a weak central lobe; (8–10 mm)   *Cerceris ruficornis*
–   Hair-tufts on 6th ventral plate golden or golden brown, not divergent, weaker, less easily seen from above (so view from side) (V.10); lower edge of clypeus with no central lobe; (6–8 mm)   ·*Cerceris quinquefasciata*

V.9

V.10

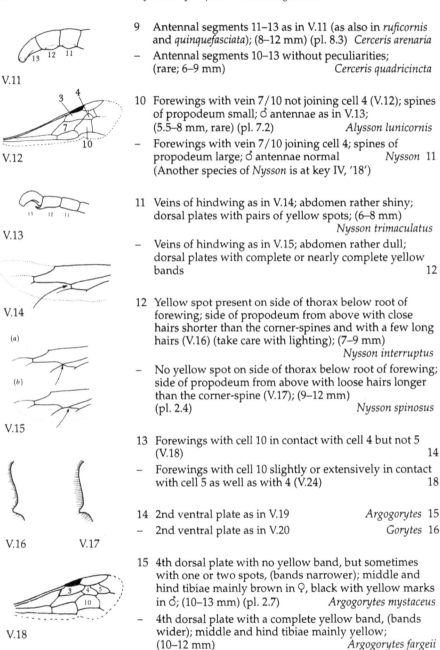

9   Antennal segments 11–13 as in V.11 (as also in *ruficornis* and *quinquefasciata*); (8–12 mm) (pl. 8.3) *Cerceris arenaria*
-   Antennal segments 10–13 without peculiarities; (rare; 6–9 mm) *Cerceris quadricincta*

V.11

10 Forewings with vein 7/10 not joining cell 4 (V.12); spines of propodeum small; ♂ antennae as in V.13; (5.5–8 mm, rare) (pl. 7.2) *Alysson lunicornis*
-   Forewings with vein 7/10 joining cell 4; spines of propodeum large; ♂ antennae normal *Nysson* 11
(Another species of *Nysson* is at key IV, '18')

V.12

11 Veins of hindwing as in V.14; abdomen rather shiny; dorsal plates with pairs of yellow spots; (6–8 mm) *Nysson trimaculatus*
-   Veins of hindwing as in V.15; abdomen rather dull; dorsal plates with complete or nearly complete yellow bands   12

V.13

12 Yellow spot present on side of thorax below root of forewing; side of propodeum from above with close hairs shorter than the corner-spines and with a few long hairs (V.16) (take care with lighting); (7–9 mm) *Nysson interruptus*
-   No yellow spot on side of thorax below root of forewing; side of propodeum from above with loose hairs longer than the corner-spine (V.17); (9–12 mm) (pl. 2.4) *Nysson spinosus*

V.14

(a)

(b)

V.15

13 Forewings with cell 10 in contact with cell 4 but not 5 (V.18)   14
-   Forewings with cell 10 slightly or extensively in contact with cell 5 as well as with 4 (V.24)   18

14 2nd ventral plate as in V.19 *Argogorytes* 15
-   2nd ventral plate as in V.20 *Gorytes* 16

V.16   V.17

15 4th dorsal plate with no yellow band, but sometimes with one or two spots, (bands narrower); middle and hind tibiae mainly brown in ♀, black with yellow marks in ♂; (10–13 mm) (pl. 2.7) *Argogorytes mystaceus*
-   4th dorsal plate with a complete yellow band, (bands wider); middle and hind tibiae mainly yellow; (10–12 mm) *Argogorytes fargeii*

V.18

16 Front of abdomen shaped as in V.21, (a yellow band occupying nearly half the 2nd dorsal plate); (7–10 mm) (pl. 8.4) *Gorytes bicinctus*
-   Front of abdomen otherwise   17

V.19   V.20

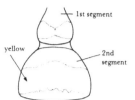

V.21

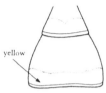

V.22

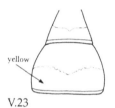

V.23

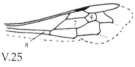

V.24

V.25

V.26

17 Front of abdomen shaped as in V.22; a yellow band occupying about ¼ of the 2nd dorsal plate; face with small yellow patches; (9–11 mm) (pl. 2.6)
   Gorytes quadrifasciatus
– Front of abdomen shaped as in V.23; a yellow band occupying not less than ⅓ of the 2nd dorsal plate; face largely yellow; (rare; 9–13 mm)   *Gorytes laticinctus*

18 Forewings with cell 7 not in contact with cell 4 (V.24); 1st abdominal segment much narrower than 2nd, bulbous at rear   *Mellinus* 19
– Forewings with cell 7 in contact with cell 4 (V.25 and V.26); 1st abdominal segment not much narrower than 2nd, enlarging evenly towards rear   20

19 Femora yellow and black; body markings bright yellow, variable; (7–14 mm) (pl. 3.1)   *Mellinus arvensis*
– Femora brown and black; body markings white; (? extinct; 6–11 mm)   *Mellinus crabroneus*

20 Boundary of eyes in face view notched (key I) (7–10 mm) (pl. 7.8) (superfamily VESPOIDEA: family Sapygidae)   *Sapyga clavicornis*
   (Another species of *Sapyga* is at key IV, '16– ')
– Boundary of eyes in face view not notched   21

21 Forewings with cell 7 not quite meeting cell 8 (V.25); (antennae obviously thickened); (8–17 mm) (pl. 8.5)
   Philanthus triangulum
– Forewings with cell 7 in contact with cell 8 (V.26)
   [superfamily APOIDEA: family Apidae (bees),
   Epeolus, *Nomada* (pl. 3.3)]

VI Solitary wasps with two submarginal cells
    (superfamilies APOIDEA and VESPOIDEA)

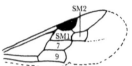

1   Forewings with cell 10 missing (VI.1)   2
–   Forewings with cell 10 present (VI.3)   4

VI.1

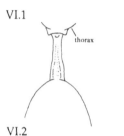

2   Front of abdomen in the form of a stalk (VI.2)
                                              Stigmus 3
–   Abdomen rounded at front; (not more than 3.5 mm)
    (pl.ˑ5.5)   Spilomena
    (Four species; no key provided - consult an expert or a
    more advanced work such as Richards (1980) or the
    revision by Dollfuss (1991))

VI.2

3   Part of mesopleuron below wing-base wrinkled; knob
    just in front of this (humeral tubercle) white; (3–5.5 mm)
                                            Stigmus solskyi
–   Part of mesopleuron below wing-base smooth and shiny
    with scattered pits; humeral tubercle black; (3.5–5.5mm)
                                           Stigmus pendulus

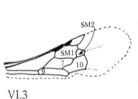

4   Forewings with cell SM1 in contact at front with
VI.3   terminal veinless area (VI.3); (♀ usually red on front of
    abdomen)   Miscophus 5
–   Forewings with cell SM1 not in contact with terminal
    veinless area (VI.4)   6

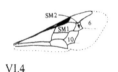

5   ♀ front of abdomen red; ♂ forehead finely pitted but
VI.4   with surface shining between the pits; (♂ black);
    (3.3–5 mm) (pl. 1.7)   Miscophus concolor
–   ♀ front of abdomen black; ♂ forehead very closely pitted
    and wrinkled; (♂ black); (3.7–5 mm)   Miscophus ater

6   Forewings with cells SM1 and SM2 about equal in their
VI.5   longest dimension (VI.5); tongue very long
    [superfamily APOIDEA: family Apidae (bees), Chelostoma]
–   Forewings with cells SM1 and SM2 very unequal (VI.4);
    tongue very short   7

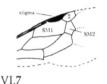

7   Front of abdomen red; yellowish white spots present on
VI.6   abdomen; forewings with cell 2 about same size as
    stigma (VI.7); (♀ antennae long, reaching to about the
    stigma); ♂ antennae as in VI.6; (extinct; 7–8 mm)
    (pl. 1.6)   Dinetus pictus
–   Front of abdomen not red; no yellowish white spots on
    abdomen; forewings with cell 2 much larger than
    stigma; ♂ antennae otherwise   8

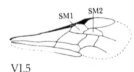

VI.7

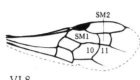

VI.8

8 Antennae nearly as long as the wings, all segments except the first two 3–4 times as long as broad; (insect black; abdomen very slender but not stalked); (male, ♀ being wingless, see key II, '6'; 7–10 mm) (superfamily VESPOIDEA: family Tiphiidae) *Methocha ichneumonides*

– Antennae shorter, not reaching nearly to the stigma of the forewings, most of the segments 2½ times as long as broad or less 9

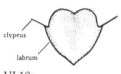

VI.9

9 Forewings with vein 10/11 joining cell SM1 or meeting vein SM1/SM2 (VI.8), or joining cell SM2 before the middle (VI.11) *Pemphredon* 10 (*In* Pemphredon *the surface sculpture varies much within the species*)

– Forewings with vein 10/11 joining cell SM2 at or beyond the middle (VI.19, VI.21) or meeting vein SM2/6 19

VI.10a

10 Stalk of abdomen not more than ⅓ as long as swollen part of 1st segment; with a horn between the antennae (VI.9); (♀ clypeus with the notch as in VI.10a or b); (4.5–8 mm) *Pemphredon morio*

– Stalk of abdomen about ⅔ as long as swollen part of 1st segment; face with no horn 11

VI.10b

11 Forewings with vein 10/11 running into cell SM2 (VI.11); (propodeum wrinkled all over); (7.5–12 mm) (pl. 5.6) *Pemphredon lugubris*

– Forewings with vein 10/11 meeting vein SM1/SM2 (VI.8) or running into cell SM1; (♀ propodeum above usually with a polished zone near the edge (VI.16)) 12

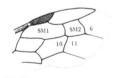

VI.11

12 Male 13
– Female 16

VI.12

13 Antennal segments all without a low thin keel (tyloid) along back; (5–6.5 mm) *Pemphredon inornata*
– Some antennal segments with a distinct tyloid 14

14 Pits of the mesonotum about half as wide as the front ocellus, flat-bottomed; (tyloids present on antennal segments 6–11); (5.5–6 mm) *Pemphredon austriaca*
– Pits of the mesonotum finer 15

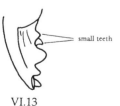

VI.13

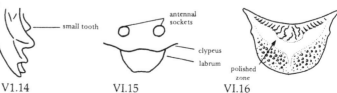

VI.14   VI.15   VI.16

VI.17

15 Dorsal area of propodeum with a polished peripheral
     zone (VI.16); (5–6 mm)   *Pemphredon lethifera*
  – Dorsal area of propodeum with a narrow, more or less
     wrinkled, strip corresponding to the polished zone of
     P. lethifera; (8–9.5 mm)   *Pemphredon rugifera*

16 Length 7.5–10.5 mm; pygidium with large, deep pits, the
     tip usually with a cluster of them and sometimes a ridge
     along the middle; (clypeus with a shallow to deep notch,
     VI.18a or b; rare)   *Pemphredon rugifera*
  – Length 5–7.5 mm; pygidium with scattered, small pits
                                                              17

VI.18a

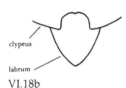

VI.18b

17 Clypeus with lower edge raised in the middle, shaped as
     in VI.12: mandibles with 4 main and 2 small teeth
     (VI.13); mesonotum usually almost unpitted; (sides and
     rear of propodeum outside the polished zone (see VI.16)
     with a close network of sharp wrinkles); (6–8 mm)
                                         Pemphredon inornata
  – Clypeus with lower edge flat and with no projection or
     with 3 central minute blunt points; mandibles with 4
     main teeth and 1 small one (VI.14); mesonotum
     distinctly pitted   18

VI.19

VI.20

18 Clypeus with lower edge straight or slightly curved
     (VI.15) (view the face from in front; do not tip it back);
     propodeum outside the polished zone pitted and with
     weak, mostly parallel, wrinkles visible only at certain
     angles; mesonotum finely and sparsely punctured;
     (5–7 mm)   *Pemphredon lethifera*
  – Clypeus with 3 central minute blunt points (VI.17) (view
     as in lead '18'); propodeum outside the polished zone
     almost as in *P. inornata* (lead '17'); punctures of
     mesonotum large, flat-bottomed; (rare; 5–7.5 mm)
                                         Pemphredon austriaca

19 Forewings with vein 7/10 meeting or almost meeting
     vein SM1/SM2 (VI.19)
             [superfamily APOIDEA: family Apidae (bees),
                                         Hylaeus (pl. 5.3)]
  – Forewings with vein 7/10 joining cell SM1 well short of
     its boundary with SM2 (VI.21)   20

VI.21

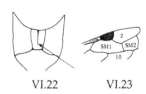

VI.22   VI.23

20 Forewings with cell SM2 somewhat shorter from front to back than from side to side (VI.23 and VI.25); front of abdomen as in VI.20; (tegulae very large)
      (superfamily VESPOIDEA: family Tiphiidae)
      Tiphia 21
      (As an aberration SM1 and SM2 may be merged)
   – Forewings with cell SM2 longer from front to back than from side to side (VI.21); front of abdomen not as in VI.20   22

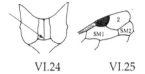

VI.24   VI.25

21 Middle ridge on propodeum usually as in VI.22; 2nd dorsal plate heavily pitted; ♂ forewings with cell 2 as in VI.23; ♀ legs mainly brownish; (♂ 5–11 mm, ♀ 7–14 mm) (pl. 5.7)   *Tiphia femorata*
   – Middle ridge on propodeum usually as in VI.24; 2nd dorsal plate scarcely at all pitted; ♂ forewings with cell 2 as in VI.25; ♂ legs mainly black; (♂ 4–6 mm, ♀ 5–6 mm)
      Tiphia minuta

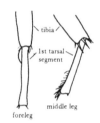

VI.26

22 Hind tibia with small spines along its length; side of thorax below wing (mesopleuron) irregularly roughened; face wide (pl. 5.4)   *Diodontus* 23
   – Hind tibia without spines along its length; mesopleuron smooth apart from 2 or 3 furrows; face roundish (pl. 6.1)
      Passaloecus 26

VI.27

23 Mandibles not yellow; first tarsal segments straight, unthickened (VI. 26)   24
   – Mandibles mainly yellow; first tarsal segments otherwise (VI.27, VI.28)   25

24 Head coarsely pitted (and wrinkled in ♂), moderately shining; ♀ front tibiae black; ♂ usually with white spot below base of forewing; (4–6.5 mm)   *Diodontus tristis*
   – Head finely pitted, dull; ♀ front tibiae yellow along front; ♂ without white spot; (4–6.5 mm)   *Diodontus luperus*

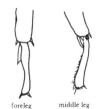

VI.28

25 ♀ face with the pits on the surface fine and close; ♀ top of propodeum at sides with an area of coarse, uneven ribbing, not much contrasted with the wrinkled central area; ♂ legs VI.27; (3–6 mm) (pl. 5.4)   *Diodontus minutus*
   – ♀ face with the pits on the surface well spaced; ♀ top of propodeum at sides with an area of fine, close ribbing, strongly contrasted with the central area; ♂ legs VI.28; (3–5 mm)   *Diodontus insidiosus*

VI.29

VI.30

VI.31

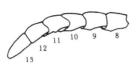

VI.32

VI.33

26 Side of thorax under wings with two distinct horizontal grooves both usually with cross-partitions (VI.29), or sometimes the upper one simple   27
   –   Side of thorax under wings with one horizontal groove containing partitions, the upper groove missing or faint and not or scarcely partitioned (VI.30)   30

27 Side of thorax with the surface in front of the vertical partitioned furrow as smooth as that behind it; clypeus with projecting middle part 3-toothed   29
   –   Side of thorax in front of the vertical partitioned furrow with wrinkles forming a more or less distinct second partitioned furrow (VI.31); clypeus with projecting middle part not toothed; (horn between antennae small and difficult to see; mandibles 2-toothed)   28

28 The 2 lengthwise furrows at front of mesonotum clearly divided by cross-bars; mesonotum with short furrows at sides in front of and behind wings, these continue as a row of pits across rear border; ♂ swellings on antennal segments (tyloids) short and prominent; (3.3–5.5 mm)
                              Passaloecus gracilis
   –   The 2 lengthwise furrows at front of mesonotum weak and not or indistinctly divided by cross-bars; mesonotum almost without pits and furrows around the edges; ♂ tyloids nearly as long as the segments, shallow; (4–5.5 mm)   *Passaloecus turionum*

29 Horn between antennae large and easy to see; ♂ antennae seen from above as in VI.32 near the tip; (4.5–6.5 mm) (pl. 6.1)   *Passaloecus corniger*
   –   Horn between antennae small, hard to see; ♂ antennae with straight-ended segments; (3.5–6.5 mm)
                              Passaloecus eremita

30 Mesonotum without lengthwise wrinkles at rear; 2nd abdominal segment constricted at rear; ♂ antennae not like a string of beads, joints with pad-like swellings (VI.33)   31
   –   Mesonotum with small lengthwise wrinkles at rear; 2nd abdominal segment not constricted; ♂ antennae like a string of beads, joints with keel-like swellings   33

31 ♀ clypeus shallowly 3-toothed; ♂ antenna with its last well-developed swelling on segment 12; (3.5–6.5 mm)
                              Passaloecus eremita
   –   ♀ clypeus with central part of margin not toothed (VI.34, 35); ♂ antenna with its last well-developed swelling on or before segment 10   32

VI.34

VI.35

VI.36

VI.37

32 Boss below base of forewing (VI.29) dark, or pale only at the rear; ♀ clypeus with lower edge narrowly turned up (illuminate from direction of lower edge) as in VI.34, and the centre only slightly swollen; ♂ antenna with its last well-developed swelling (VI.33) on segment 9 or 10; (3.5–5.5 mm)   *Passaloecus singularis*

– Boss below base of forewing pale; ♀ clypeus with lower edge broadly turned up and the centre strongly swollen (VI.35); ♂ antenna with its last well-developed swelling on segment 8; (3–5 mm)   *Passaloecus clypealis*

33 ♀ top of head sunken on the side of each rear ocellus nearest the eye; ♀ labrum usually pale; ♂ antenna with its last ridge on segment 10, the segments less swollen (VI.36); (southern species; 4.5–6 mm) *Passaloecus insignis*

– ♀ top of head not sunken between rear ocelli and the eyes (though the area between the 3 ocelli is elevated); ♀ labrum brown; ♂ antenna with its last ridge on segment 11, the segments more swollen (VI.37); (northern species; 4.5–6 mm)   *Passaloecus monilicornis*

VII Solitary wasps with one submarginal cell (pl. 4) (superfamily APOIDEA and aberrant VESPOIDEA)

VII.1

boundary

VII.2

VII.3   VII.4

VII.5

1   Forewings with cell 10 complete (VI.21)
                    Tiphia (aberrant), key VI, '21'
–   Forewings with cell 10 incomplete or absent (VII.7)   2

2   Boundary of eyes notched (VII.1)   *Trypoxylon* 3
–   Boundary of eyes not notched   7

3   Front tibiae and tarsi brownish or dirty yellow on inner
    side; top of propodeum with a clear boundary (VII.2); ♂
    antennae distinctly club-shaped; (4.5–8 mm)
                              Trypoxylon clavicerum
–   Front tibiae and tarsi black or blackish brown; top of
    propodeum with boundary faint or none; ♂ antennae
    only slightly thickened towards the tip   4

4   1st abdominal segment very narrow (VII.3); (7–11 mm)
                              Trypoxylon attenuatum
–   1st abdominal segment wider (VII.4); (5–12 mm) (pl. 6.2)
                                                           5

5   ♀ lower edge of clypeus with uniformly curved bay on
    either side of central lobe (VII.5); ♀ pit in hind coxa
    beneath (almost circular: use high magnification) with
    hairs forming a gully-like structure (use very high
    magnification); ♂ length of last antennal segment 2–2.2
    times its basal width, and of next-to-last segment
    0.75–0.90 times its width; (♀ 6.5–12 mm, ♂ 6–8.5 mm)
                              Trypoxylon medium
–   ♀ lower edge of clypeus on either side of central
    projection straight or slightly sinuous; ♀ hairs around
    the pit on the hind coxa not forming a gully-like
    structure; ♂ length of last antennal segment 2.2–3.6 times
    its basal width and of next-to-last segment 0.5–0.8 times
    its width   6

6   Transverse flange below thorax behind sockets of
    forelegs not drawn out into a backwardly curved point,
    though somewhat heightened at the centre; ♀ pit in hind
    coxa beneath circular (use high magnification);
    ♀ 6–9mm; ♂ 5–7.5mm   *Trypoxylon minus*
–   Transverse flange below thorax behind sockets of
    forelegs drawn out into a backwardly curved central
    point; ♀ pit on hind coxa elongate, rarely circular (use
    high magnification); ♀ 9–12mm; ♂ 7.5–10mm (pl. 6.2)
                              Trypoxylon figulus

VII.6

7 Forewings with vein 7/SM indistinct (VII.6);
(propodeum with a large curved spine; scutellum with
2 translucent flanges)   *Oxybelus* 8
– Forewings with vein 7/SM distinct   10

8 Middle femora in ♀ light brown, in ♂ partly yellow;
spine on thorax pointed; with glistening silvery hairs
pressed to the body (in ♂ best seen on mesothorax with
light from front); (5–9 mm)   *Oxybelus argentatus*
– Middle femora mainly black or dark brown; spine on
thorax rounded at tip; body hairs otherwise   9

9 Mandibles dark brown or black; ♀ sides of thorax below
wings with a heavy network of wrinkles; ♂ 2nd–5th
ventral plates without dense hair-fringes seen in surface
view; (5–8 mm) (pl. 3.7)   *Oxybelus uniglumis*
– Mandibles partly pale yellow and pale brown; ♀ sides of
thorax below wings smooth and shiny with distinct pits;
♂ 2nd–5th ventral plates with dense fringes of fine hairs
emerging from beneath the translucent rear margin of
the segment in front; (5–7.5 mm)
                              Oxybelus mandibularis

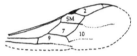

VII.7

10 Forewings with cell 9 incomplete at far end (VII.7);
borders of eyes diverging from level of front ocellus to
level of antennae; (insect 3–4.5 mm)   *Nitela* 11
– Forewings with cell 9 complete; borders of eyes
converging from level of front ocellus to level of
antennae (family Sphecidae, subfamily Crabroninae) 12

11 Propodeum with a coarse net-like pattern, shiny; first
two dorsal plates scarcely pitted; ♂ clypeus deeply 3-
lobed; (3–4.5 mm)   *Nitela borealis*
– Propodeum with a fine net-like pattern, dull; first two
dorsal plates distinctly and rather densely pitted; ♂
clypeus shallowly 3-lobed; (3–4 mm)
                    Nitela species [see Alphabetical Checklist]

VII.8

12 Front of abdomen in the form of a stalk (petiole),
1st segment as in VII.8   *Rhopalum* 13
– Front of abdomen not petioled, 1st segment not as in
VII.8   15

13 Abdomen red or reddish towards front; ♀ clypeus with
the central lobe small (blunt); ♂ antennae simple;
(4–6 mm) (pl. 4.7)   *Rhopalum clavipes*
– Abdomen with no red; ♀ clypeus with the central lobe
large (blunt or pointed); ♂ antennal segments 2 and 4
lobed (segment 3 minute)   14

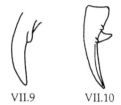

VII.9   VII.10

VII.11

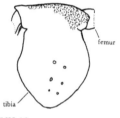

femur

tibia

VII.12

VII.13

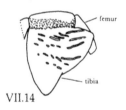

femur

tibia

VII.14

VII.15

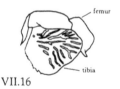

femur

tibia

VII.16

14 Clypeus with 3 points; ♀ hind tibiae pale brown towards
     tip; ♂ antennae yellow-banded near tip; (5–7 mm)
                                             Rhopalum coarctatum
  – Clypeus with a large central blunt lobe and a small point
     on either side; ♀ hind tibiae blackish brown at tip; ♂
     antennae not yellow-banded; (4.5–6 mm)
                                             Rhopalum gracile

15 Eyes beset with short hairs (3.5–5.5 mm) (pl. 6.3)
                                             Entomognathus brevis
  – Eyes without hairs   16

16 Ocelli forming an obtuse-angled triangle   17
  – Ocelli forming a right-angled or equilateral triangle   22

17 Mandibles with the tip simply pointed (VII.9 & 10); body
     mainly black   *Lindenius* 18
  – Mandibles with 2–4 teeth at the tip; body with
     conspicuous yellow markings   19

18 Side of thorax below wings with sparse short hairs;
     mandibles as in VII.9; (5–8 mm) (pl. 6.4)
                                             Lindenius albilabris
  – Side of thorax below wings with dense long hairs,
     especially ventrally; mandibles as in VII.10;
     (4.5–7.5 mm)   *Lindenius panzeri*

19 Abdomen strongly pitted; ♀ only (6 abdominal
     segments; ♂ is in couplet '22'); (?extinct)
                                             Lestica clypeata
  – Abdomen polished, pits fine and indistinct   20

20 Front tibia of ♂ (count 7 abdominal segments) with
     shield-like extension; ♀ pygidium flat; (♂ antennal
     segments 13, those towards the middle progressively
     enlarged)   *Crabro* 21
  – Front tibia of ♂ (count 7 abdominal segments) not much
     enlarged; ♀ pygidium gutter-shaped at tip; (♂ antennal
     segments 12, sometimes unequal, knobbly and hard to
     count)   *Ectemnius* 26

21 Propodeum and ♂ front tibia as in:
     VII.11 and VII.12; (10–15 mm)
     (pls. 4.1 and 4.2)   *Crabro cribrarius*
     VII.13 and VII.14; (9–13 mm)   *Crabro peltarius*
     VII.15 and VII.16; (7–11 mm)   *Crabro scutellatus*

VII.17

VII.18   VII.19

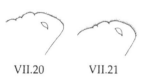

VII.20   VII.21

VII.22

VII.23

22 Head from above appearing stalked; (♂ only, but
    antennal segments 12; ♀ is in couplet '19');
    (?extinct; 8–11 mm) (pl. 8.6)   *Lestica clypeata*
 – Head appearing normal from above   23

23 Side view of head with projection as in VII.17 or VII.18,
    not as in VII.19; (body with variable yellow pattern,
    rarely without; ♂ mandibles pointed, ♀ 2-toothed);
    (5–10 mm) (pls. 4.5 and 4.6)
            Crossocerus (subgenus *Hoplocrabro*) *quadrimaculatus*
 – No such point on side of head (VII.19)   24

24 Abdomen with a pattern of yellow marks   25
 – Abdomen without yellow marks   37

25 Thorax everywhere very dull with dense surface
    sculpture; ♂ with knobs and hollows on antennae
                                                Ectemnius 26
 – Thorax with shiny areas; ♂ antennae normal
        Crossocerus subgenera *Acanthocrabro* and *Cuphopterus* 35

26 Mesonotum with the hairs short (VII.20)   27
 – Mesonotum with the hairs long (VII.21)   28

27 Pronotum with spines (VII.22), and marked with yellow;
    ♀ hairs on centre of clypeus slightly golden; ♂ 5th
    antennal segment not hollowed; (6.5–10.5 mm)
                                            Ectemnius dives
 – Pronotum without spines (VII.23), not or scarcely
    marked with yellow; ♀ hairs on centre of clypeus silvery;
    ♂ 5th antennal segment hollowed; (6–9 mm)
                                          Ectemnius borealis

28 ♂ antennae conspicuously knobbly, 3rd segment with a
    tooth halfway along it; ♀ hairs on clypeus golden; ♀ top
    of head without a definite oval depression alongside
    each eye   29
 – ♂ antennae not knobbly, almost regular, 3rd segment
    without a tooth at the middle; ♀ hairs on clypeus silvery
    or only faintly golden; ♀ top of head with a distinct oval
    depression, usually shining, alongside each eye (view
    from above and tilt backwards slightly)   32

VII.24

VII.25

VII.26

VII.27

VII.28

VII.29

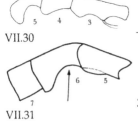

VII.30

VII.31

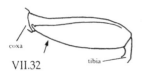

VII.32

29 ♀ propodeum with a curved ridge clearly separating the sculpture of the rear surface from that of the upper; ♂ antennae as in VII.24; (9–12 mm) (pl. 4.3)
    Ectemnius lapidarius

–   ♀ propodeum without a ridge separating the rear and upper surfaces; ♂ antennae as in VII.26, VII.28 or VII.30
    30

30 ♀ clypeus as in VII.25 (view with mandibles towards the light; lateral teeth divergent, far from the narrow central lobe); ♂ antennae as in VII.26; (rare; 10–14 mm)
    Ectemnius ruficornis

–   ♀ clypeus as in VII.27 or VII.29 (view with mandibles towards the light; lateral teeth scarcely divergent); ♂ antennae otherwise   31

31 ♀ clypeus as in VII.27 (lateral teeth close to wide central lobe); ♀ ventral plates without yellow spots; ♂ antennae as in VII.28; (11–16.5 mm)   *Ectemnius cavifrons*

–   ♀ clypeus as in VII.29 (lateral teeth rather far from narrow central lobe); ♀ ventral plates with yellow spots; ♂ antennae as in VII.30, 3rd segment with hairs on tooth; (10–17 mm)   *Ectemnius sexcinctus*

32 Pronotum rounded at the corners, sloping down towards the front; ♂ antennae with 6th segment not notched   33

–   Pronotum sharply angled, or with a small tooth at each corner, scarcely sloping down towards the front; ♂ antennae notched as in VII.31   34

33 Mesonotum almost without pits but with bold wrinkles, transverse at the front, lengthwise at the rear; hairs of 1st dorsal plate long and erect; (9–17 mm)
    Ectemnius cephalotes

–   Mesonotum with pits set in a network of fine lengthwise and faint transverse wrinkles; hairs of 1st dorsal plate short and leaning backwards; (9–14.5 mm)
    Ectemnius lituratus

34 ♀ rear of propodeum with horizontal wrinkles continuous with those at the sides; ♂ front femur as in VII.32; (8–14.5 mm) (pl. 4.4)   *Ectemnius continuus*

–   ♀ rear of propodeum very rough, horizontal wrinkles scarcely evident; ♂ front femur as in VII.33; (6–9.5 mm)
    Ectemnius rubicola

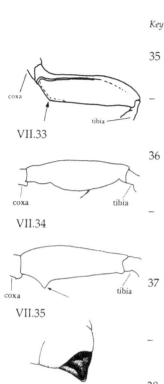

VII.33

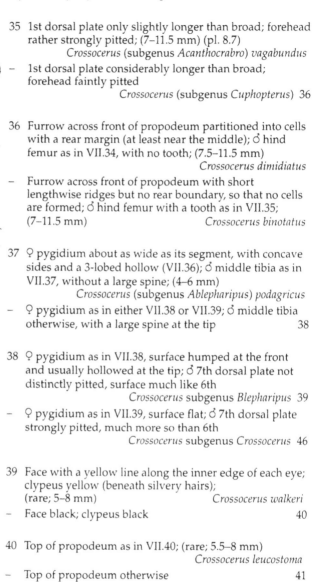

35 1st dorsal plate only slightly longer than broad; forehead rather strongly pitted; (7–11.5 mm) (pl. 8.7)
Crossocerus (subgenus *Acanthocrabro*) *vagabundus*
– 1st dorsal plate considerably longer than broad; forehead faintly pitted
Crossocerus (subgenus *Cuphopterus*) 36

VII.34

36 Furrow across front of propodeum partitioned into cells with a rear margin (at least near the middle); ♂ hind femur as in VII.34, with no tooth; (7.5–11.5 mm)
Crossocerus dimidiatus
– Furrow across front of propodeum with short lengthwise ridges but no rear boundary, so that no cells are formed; ♂ hind femur with a tooth as in VII.35; (7–11.5 mm)
Crossocerus binotatus

VII.35

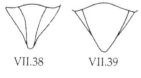

VII.36

37 ♀ pygidium about as wide as its segment, with concave sides and a 3-lobed hollow (VII.36); ♂ middle tibia as in VII.37, without a large spine; (4–6 mm)
Crossocerus (subgenus *Ablepharipus*) *podagricus*
– ♀ pygidium as in either VII.38 or VII.39; ♂ middle tibia otherwise, with a large spine at the tip 38

38 ♀ pygidium as in VII.38, surface humped at the front and usually hollowed at the tip; ♂ 7th dorsal plate not distinctly pitted, surface much like 6th
Crossocerus subgenus *Blepharipus* 39
– ♀ pygidium as in VII.39, surface flat; ♂ 7th dorsal plate strongly pitted, much more so than 6th
Crossocerus subgenus *Crossocerus* 46

VII.37

39 Face with a yellow line along the inner edge of each eye; clypeus yellow (beneath silvery hairs); (rare; 5–8 mm) *Crossocerus walkeri*
– Face black; clypeus black 40

VII.38   VII.39

40 Top of propodeum as in VII.40; (rare; 5.5–8 mm)
Crossocerus leucostoma
– Top of propodeum otherwise 41

VII.40

41 Hairs on top of head and mesonotum long; (♂ tibiae and tarsi without noticeable enlargements) 42
– Hairs on top of head and mesonotum short, either sparse or forming a close but inconspicuous pile; (♂ tibiae or tarsi sometimes with noticeable enlargements) 43

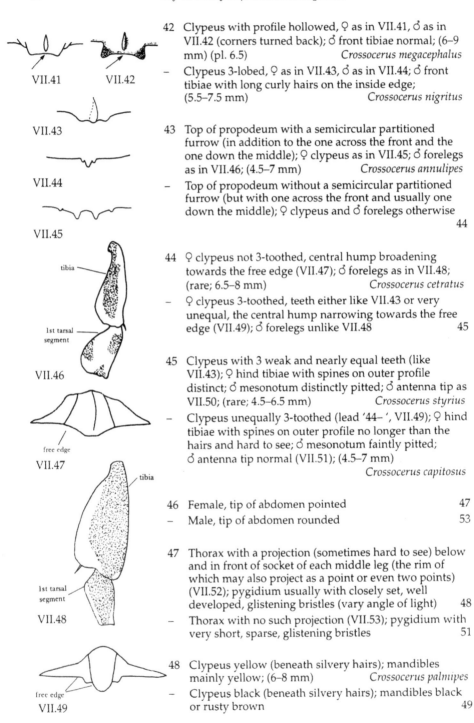

VII.41   VII.42

VII.43

VII.44

VII.45

tibia

1st tarsal
segment

VII.46

free edge

VII.47

tibia

1st tarsal
segment

VII.48

free edge

VII.49

42 Clypeus with profile hollowed, ♀ as in VII.41, ♂ as in
    VII.42 (corners turned back); ♂ front tibiae normal; (6–9
    mm) (pl. 6.5)   *Crossocerus megacephalus*
–   Clypeus 3-lobed, ♀ as in VII.43, ♂ as in VII.44; ♂ front
    tibiae with long curly hairs on the inside edge;
    (5.5–7.5 mm)   *Crossocerus nigritus*

43 Top of propodeum with a semicircular partitioned
    furrow (in addition to the one across the front and the
    one down the middle); ♀ clypeus as in VII.45; ♂ forelegs
    as in VII.46; (4.5–7 mm)   *Crossocerus annulipes*
–   Top of propodeum without a semicircular partitioned
    furrow (but with one across the front and usually one
    down the middle); ♀ clypeus and ♂ forelegs otherwise
                                                            44

44 ♀ clypeus not 3-toothed, central hump broadening
    towards the free edge (VII.47); ♂ forelegs as in VII.48;
    (rare; 6.5–8 mm)   *Crossocerus cetratus*
–   ♀ clypeus 3-toothed, teeth either like VII.43 or very
    unequal, the central hump narrowing towards the free
    edge (VII.49); ♂ forelegs unlike VII.48   45

45 Clypeus with 3 weak and nearly equal teeth (like
    VII.43); ♀ hind tibiae with spines on outer profile
    distinct; ♂ mesonotum distinctly pitted; ♂ antenna tip as
    VII.50; (rare; 4.5–6.5 mm)   *Crossocerus styrius*
–   Clypeus unequally 3-toothed (lead '44– ', VII.49); ♀ hind
    tibiae with spines on outer profile no longer than the
    hairs and hard to see; ♂ mesonotum faintly pitted; ♂
    antenna tip normal (VII.51); (4.5–7 mm)
                                             Crossocerus capitosus

46 Female, tip of abdomen pointed   47
–   Male, tip of abdomen rounded   53

47 Thorax with a projection (sometimes hard to see) below
    and in front of socket of each middle leg (the rim of
    which may also project as a point or even two points)
    (VII.52); pygidium usually with closely set, well
    developed, glistening bristles (vary angle of light)   48
–   Thorax with no such projection (VII.53); pygidium with
    very short, sparse, glistening bristles   51

48 Clypeus yellow (beneath silvery hairs); mandibles
    mainly yellow; (6–8 mm)   *Crossocerus palmipes*
–   Clypeus black (beneath silvery hairs); mandibles black
    or rusty brown   49

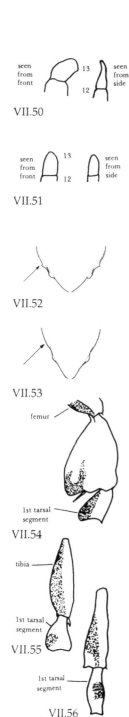

seen from front 13

seen from side 12

VII.50

seen from front 13

seen from side 12

VII.51

VII.52

VII.53

femur

1st tarsal segment

VII.54

tibia

1st tarsal segment

VII.55

1st tarsal segment

VII.56

49 Top of propodeum polished on either side of the middle furrow, the encircling furrow distinct and complete; pygidium red at the tip for nearly ⅓ of its length; (4.5–6 mm)   *Crossocerus ovalis*

– Top of propodeum with close, fine, parallel ribbing on either side of the middle furrow, not polished here though not so dull as adjacent areas, the encircling furrow indistinct at the sides; pygidium black or slightly reddish at extreme tip   50

50 (*The females in this couplet are difficult to distinguish*)
Middle furrow of propodeum boat-shaped, widest at middle or rear, more distinct from front furrow; mesonotum with pits sparser, rather shiny between them; (scutellum usually without a yellow spot); (5–6.5 mm)   *Crossocerus tarsatus*

– Middle furrow of propodeum tapering strongly and progressively from near the front, less distinct from front furrow; mesonotum with pits closer, rather dull between them; (scutellum sometimes with a yellow spot); (5.5–6.5 mm)   *Crossocerus pusillus*

51 Rear (sunken) border of mesonotum with fine lengthwise ridges (put the head towards you and tilt the insect forwards)   52

– Rear (sunken) border of mesonotum without fine lengthwise ridges (view as indicated); (thorax without yellow marks; middle tibiae largely yellow); (4–5 mm)   *Crossocerus exiguus*

52 Thorax with yellow marks; middle tibiae largely yellow; tip of pygidium obviously rusty red; (5–6 mm) (pl. 6.7)   *Crossocerus wesmaeli*

– Thorax usually without yellow marks; middle tibiae mainly black; tip of pygidium black or hardly at all reddish
    common; 5–6.5 mm; pl. 6.6:   *Crossocerus elongatulus*
    rare, not satisfactorily distinguishable in ♀ sex;
    5–6.5 mm: *Crossocerus distinguendus*

53 Forelegs as VII.54; (5–6.5 mm)   *Crossocerus palmipes*
– Forelegs as VII.55; (4–4.5 mm)   *Crossocerus tarsatus*
– Forelegs as VII.56; (4.5–5.5 mm)   *Crossocerus pusillus*
– Forelegs slimmer, tarsus dark   54

54 Clypeus yellow; (3.5–4 mm)   *Crossocerus exiguus*
– Clypeus black   55

55	Pygidium present (VII.57); (4–4.5 mm) *Crossocerus ovalis*
–	Pygidium absent; (VII.58)					56

56	Middle femur as VII.59; (5–6 mm)
						Crossocerus elongatulus
–	Middle femur as VII.60					57

VII.57

57	Front femur without long hair fringe below; (4–4.5 mm)
						Crossocerus wesmaeli
–	Front femur with long hair fringe below;
	(rare; 4–5.5 mm)			*Crossocerus distinguendus*

VII.58

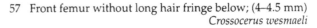

VII.59

VII.60

4 Techniques and approaches to original work

Catching and handling wasps for identification

Fig. 22. A pinned wasp

Occasionally a wasp can be trapped in a specimen tube as she emerges from her nest entrance, but most are too wary for this. A flying wasp can be caught with a butterfly net by sweeping deftly and then quickly flipping the end of the net over the frame to prevent the insect from flying out. Female solitary wasps can sting but rarely do. Nevertheless, treat the larger species with respect.

For critical identification it is usually necessary to kill a specimen. A wasp can be killed in a corked tube with about 1 cm of dental plaster set in the bottom (or a strip of filter paper) to absorb a killing fluid such as ethyl acetate. Specimens killed in ethyl acetate remain flexible for some time. To avoid wetting the specimen, dry the dental plaster thoroughly before use, add only a few drops of ethyl acetate, and use absorbent corks rather than plastic tube closures.

When dead, the wasp should be pinned to facilitate handling and storage. The very largest can be pinned with an entomological pin straight through the thorax (fig. 22). Smaller wasps should be pinned on a fine headless pin which in turn is 'staged' on a strip of Plastazote (expanded polyethylene foam) trimmed with a razor blade to the size of half a matchstick, held on a larger pin (fig. 23). If Plastazote is not available the traditional polyporus is almost as good. While the wasp is still flexible it should be spread-eagled on its pin to give a clear view of wing veins, legs, the rear and one side of the thorax, and the upper and lower surfaces of the front of the abdomen. If necessary, set the wasp on a sheet of cork or Plastazote, using pins to hold the bits in position for a few days until the wasp becomes rigid. If a female wasp is caught with her prey, both should be staged on the same pin.

Fig. 23. A staged wasp

Every specimen must carry on its large pin a label, giving the place and date of capture and the name of the captor, and sometimes other information or a coded reference to more detailed notes. This data label always stays with the specimen, which would be of no biological value without it. A second label bearing the wasp's name may be added lower on the pin.

A good hand-lens may reveal enough detail of the wasp's structure to enable you to use the Quick-Check Key, but the main keys require a stereoscopic dissecting microscope with a magnification of x20 or more. Use the best light source you can get, if necessary focussing the light on the specimen with a lens. Vary the angle of observation exhaustively. This is not very difficult if the wasp is pinned at various angles on an L-shaped stand improvised from Plastazote or cork. The more versatile 'hinge-and-bracket' stand (fig. 24) takes longer to make but is easier to use. A similar stand is available from entomological suppliers (p. 58).

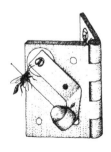

Fig. 24. A 'hinge and bracket' stand

Pinned wasps can be stored in a conventional cork-lined insect box (expensive) or in a sandwich-box with a tight-fitting floor of Plastazote. To avoid damage by insects and mites, these boxes should have close-fitting lids. The wooden ones can be treated with paradichlorbenzene or Rentokil Woodworm Fluid.

Suppliers of nets, pins, insect boxes, etc. are:
Watkins and Doncaster, Conghurst Lane, Four Throws, Hawkhurst, Kent, TN18 5ED.
Worldwide Butterflies Ltd, Compton House, nr Sherborne, Dorset DT9 4QN.

Plastazote is available (as sheets 1000 x 750 x 12 mm) *from:*
Polyformes Ltd, Cherrycourt Way, Stanbridge Road, Leighton Buzzard, Bedfordshire LU7 8UH.
It is ideal for pinning and storing insects. Expanded polystyrene ceiling tiles are a cheaper and less satisfactory alternative.

National recording scheme

Anyone who studies wasps (or bees and ants) is welcome to participate in the work of the Bees, Wasps and Ants Recording Society (BWARS), which collates records, and will publish distribution maps, for the British species. The Society will give help with identification. Enquiries should be addressed to the Honorary Secretary, Mr David Lloyd, 1 Crest Road, Rochester, Kent ME1 2NG. Queries about identification should be sent (with a stamped addressed envelope) to Robin Edwards, 5 St Edward's Close, East Grinstead, West Sussex RH19 1JP. Do not send specimens until contact has been made.

Marking wasps for experiments

For marking, wasps can be cooled in a refrigerator or anaesthetised in the field by squirting into the specimen tube a light dose of carbon dioxide from a sparklet-operated dispenser. One type is available from Customer Service Department, Edme Ltd, Mistley, Manningtree, Essex CO11 1HG, or a wine bottle opener such as a Corkette works well. The paint must be quick-drying. Hobby paints or typists' correction fluid such as Tipp-Ex are suitable. A tiny spot can be applied with a fine grass-stem, and must dry before the wasp recovers, or she will try to groom it off and gum herself up. Marked wasps require intensive care until they are fully recovered.

Studying nests

Wasp burrows in sandy soil are liable to collapse when excavated. A long flexible grass-stem pushed down the hole beforehand acts as a useful guide when one digs out a nest to locate the cells at the bottom; or plaster of Paris can be poured down a hole to make a cast of the burrow system. Nests in bramble stems can be exposed by carefully splitting the stems, but those in wood require patient excavation. The contents of the individual cells should be kept separate. It should be possible to discover the nature and sex of the prey; the position and stage of development of the young wasp; and the nature of any parasites or kleptoparasites. To rear out the wasp or its parasite (a long job if an overwintering stage intervenes) involves keeping the cell contents humid but not wet. Gelatin pharmaceutical capsules may be used as substitute cells for live larvae or pupae if the nest cannot be reassembled. If parasites can be reared to the adult stage it may be possible to identify them, using Chinery (1976) or, for flies, Unwin (1981) or Colyer and Hammond (1968), to find out what group they belong to and what keys are available. Danks (1971) gives keys to the nests and parasites of wasp species nesting in bramble stems.

A bundle of cut dead bramble stems about 30–50 cm long, set out as a trap-nest in a dry sunny place, may attract various species. The nests can be examined by splitting the stems, which can then be put back together held with elastic bands, to allow the wasps and parasites to develop.

Microclimate studies

Unwin and Corbet (1991) and Unwin (1978) explain how equipment for measuring microclimate can be built simply and cheaply, and how it can be used in biological studies. Unwin (1980) gives a more advanced treatment of the subject.

Entomological societies

Some readers may wish to join a society such as the Amateur Entomologists' Society (AES Registrar, 5 Oakfield, Plaistow, Billingshurst, West Sussex RH14 0QD), the British Entomological and Natural History Society (The Pelham Clinton Building, Dinton Pastures Country Park, Davis Street, Hurst, Reading, Berkshire RG10 0TH) or the Royal Entomological Society (41 Queens Gate, London SW7 5HU). These publish attractive journals which contain articles on entomological topics and keep members in touch with various meetings and recording schemes.

How to present the results of research

Writing up is an important part of a research project, particularly when the findings are to be communicated to other people. A really thorough, critical investigation that has established new information of general interest may be worth publishing if the animals on which it is based can be identified with certainty. Journals that publish short papers on insect biology include the Entomologist's Monthly Magazine, Entomologist's Gazette, Bulletin of the Amateur Entomologists' Society, The Entomologist, and, for material with an educational slant, Journal of Biological Education.

Those unfamiliar with publishing conventions are advised to examine current numbers of these journals to see what sort of thing they publish, and then to write a paper along similar lines, keeping it as short as possible while presenting enough information to establish the conclusions. A book on writing papers for publication, such as Day (1989), can be very helpful. It is then time to consult an appropriate expert who can give advice on whether and in what form the material might be published.

It is an unbreakable convention of scientific publication that results are reported with scrupulous honesty. Hence it is essential to keep detailed and accurate records throughout the investigation, and to distinguish in the write-up between certainty and probability, and deduction and speculation. In many cases it will be necessary to apply appropriate statistical techniques to test the significance of the findings. A book such as Chalmers and Parker's (1989) OU Project Guide will help, but this is an area where expert advice can contribute much to the planning, as well as the analysis, of the work.

Further reading

Finding books

Some of the books and journals listed here will be unavailable in local and school libraries. It is possible to make arrangements to see or borrow such works by seeking permission to visit the library of a local university, or by asking your local public library to borrow the work (or a photocopy of it) for you via the British Library, Document Supply Centre. This may take several weeks, and it is important to present your librarian with a reference that is correct in every detail. References are acceptable in the form given here, namely the author's name and date of publication, followed by (for a book) the title and publisher or (for a journal article) the title of the article, the journal title, the volume number, and the first and last pages of the article.

References

Alcock, J., Barrows, E.M., Gordh, G., Hubbard, L.J., Kirkendall, L., Pyle, D.W., Ponder, T., & Zalom, F.G. (1978). The ecology and evolution of male reproductive behaviour in the bees and wasps. *Zoological Journal of the Linnean Society*, **64**, 293–326.

Baerends, G.P. (1941). Fortpflanzungsverhalten und Orientierung der Grabwespe *Ammophila campestris* Jur. *Tijdschrift voor Entomologie*, **84**, 68–275. (For a brief account in English see Tinbergen, 1958.)

de Beaumont, J. (1964). *Hymenoptera: Sphecidae*. Insecta Helvetica 3. Lausanne: Imprimeria la Concorde.

Bohart, R.M. & Marsh, P. M. (1960). Observations on the habits of *Oxybelus sericeum* Robertson (Hymenoptera: Sphecidae). *Pan-Pacific Entomologist*, 36, 115–18.

Bohart, R.M. & Menke, A.S. (1976). *Sphecid wasps of the world*. Berkeley: University of California Press.

Brockmann, H.J. (1979). Nest-site selection in the great golden digger wasp, *Sphex ichneumoneus* L. (Sphecidae). *Ecological Entomology*, 4, 211–24.

Butler, C.G. (1965). Sex attraction in *Andrena flavipes* Panzer (Hymenoptera: Apidae), with some observations on nest-site restriction. *Proceedings of the Royal Entomological Society of London, Series A*, **40**, 77–80.

Chalmers, N. and Parker, P. (1989). *OU Project Guide: Field Work and Statistics for Ecological Projects*. London: Open University/Field Studies Council.

Chinery, M. (1976). *A Field Guide to the Insects of Britain and Northern Europe*, 2nd (corrected) edn. London: Collins.

Colyer, C.N. & Hammond, C.O. (1968). *Flies of the British Isles*. London: Frederick Warne.

Corbet, S.A. & Backhouse, M. (1975). Aphid-hunting wasps: a field study of *Passaloecus*. *Transactions of the Royal Entomological Society of London*, **127**, 11–30.

Danks, H.V. (1971). Biology of some stem-nesting aculeate Hymenoptera. *Transactions of the Royal Entomological Society of London*, **122**, 323–99.

Day, M.C. (1988). *Spider wasps (Pompilidae)*. Handbooks for the Identification of British Insects **6** (4). London: Royal Entomological Society.

Day, R.A. (1989). *How to Write and Publish a Scientific Paper*. Cambridge: Cambridge University Press.

Dollfuss, H. (1991). Bestimmungsschlüssel der Grabwespen Nord- und Zentraleuropas (Hymenoptera, Sphecidae). *Stapfia* **24**, 1–247

Eberhard, M.J.W. (1978). Polygyny and the evolution of social behaviour in wasps. *Journal of the Kansas Entomological Society*, **51**, 832–56.

Eisner, T. (1970). Chemical defense against predation in arthropods. In *Chemical Ecology*. E. Sondheimer & J.B. Simeone (editors) pp. 157–217. New York & London: Academic Press.

Evans, H.E. (1962). The evolution of prey-carrying mechanisms in wasps. *Evolution*, **16**, 468–83.

Evans, H.E. (1963). Predatory wasps. *Scientific American*, **208**, 145–54.

Evans, H.E. & Eberhard, M.J.W. (1970). *The Wasps*. Ann Arbor: University of Michigan Press.

Falk, S. (1991). *A review of the Scarce and Threatened Bees, Wasps and Ants of Great Britain*. Peterborough: Nature Conservancy Council.

Field, J. (1989). Intraspecific parasitism and nesting success in the solitary wasp *Ammophila sabulosa*. *Behaviour*, **110**, 1–4.

Field, J. (1992). Patterns of nest provisioning and parental investment in the solitary digger wasp *Ammophila sabulosa*. *Ecological Entomology*, **17**, 43–51.

Gauld, I. & Bolton, B. (1988). *The Hymenoptera*. Oxford: British Museum (Natural History) and Oxford University Press.

ITE Overlays (1978). *Overlays of Environmental and Other Factors for use with Biological Records Centre Distribution Maps*. Published by Institute of Terrestrial Ecology, 68 Hills Road, Cambridge CB2 1LA.

Iwata, K. (1942). Comparative studies on the habits of solitary wasps. *Tenthredo*, **4**, 1–146.

Lomholdt, O. (1975). *The Sphecidae (Hymenoptera) of Fennoscandia and Denmark*. Fauna Entomologica Scandinavica vol. 4, parts 1 and 2. Klampenborg, Denmark: Scandinavian Science Press.

Olberg, G. (1959). *Das Verhalten der solitären Wespen Mitteleuropas*. Berlin: VEB Deutscher Verlag der Wissenschaften.

Peckham, D.J. (1977). Reduction of miltogrammine cleptoparasitism by male *Oxybelus subulatus* (Hymenoptera: Sphecidae). *Annals of the Entomological Society of America*, **70**, 823–828.

Richards, O.W. (1978). Aculeata. In M.G. Fitton and others, *A Check List of British Insects*, 2nd edn completely revised, part 4, *Hymenoptera*. Handbooks for the Identification of British Insects XI(4). London: Royal Entomological Society of London. (Available from Royal Entomological Society, 41 Queens Gate, London SW7 5HU.)

Richards, O.W. (1980). *Scolioidea, Vespoidea and Sphecoidea (Hymenoptera Aculeata)*. Handbooks for the Identification of British Insects VI(3b). London: Royal Entomological Society of London. (Available from Royal Entomological Society, 41 Queens Gate, London SW7 5HU.)

Shuckard, W.E. (1837). *Essays on the Indigenous Fossorial Hymenoptera*. London: published by the author at 31 Robert Street, Chelsea.

Thornhill, R. & Alcock, J. (1983). *The evolution of insect mating systems*. Cambridge, Mass.: Harvard University Press.

Tinbergen, N. (1958). *Curious Naturalists*. London: Country Life.

Unwin, D.M. (1978). Simple techniques for microclimate measurement. *Journal of Biological Education*, **12**, 179–189.

Unwin, D.M. (1980). *Microclimate Measurement for Ecologists*. New York & London: Academic Press.

Unwin, D.M. (1981). A key to the families of British Diptera. *Field Studies*, **5**, 513–533.

Unwin, D.M. and Corbet, S.A. (1991). *Insects, Plants and Microclimate*. Slough: The Richmond Publishing Co. Ltd.

Willmer, P.G. (1985a). Size effects on hygrothermal balance and foraging patterns in a sphecid wasp, *Cerceris arenaria*. *Ecological Entomology*, **10**, 469–479.

Willmer, P.G. (1985b). Thermal ecology, size effects, and the origins of communal behaviour in *Cerceris* wasps. *Behavioural Ecology and Sociobiology*, **17**, 151–160.

Wilson, E.O. (1975). *Sociobiology, The New Synthesis*. Cambridge, Massachusetts: The Belknap Press of Harvard University Press.

Systematic checklist of genera

Systematic arrangement of solitary wasp genera into superfamilies (ending in -OIDEA), families (-idae) and subfamilies (-inae) (based on Gauld and Bolton (1988); Vespoidea sequence from Richards (1980); Apoidea sequence from Dollfuss (1991), based on a world treatment by Bohart & Menke (1976)).

VESPOIDEA
Sapygidae
  Sapyga
Tiphiidae
  Tiphiinae
    Tiphia
  Methochinae
    Methocha
Mutillidae
  Mutillinae
    Mutilla
    Smicromyrme
  Myrmosinae
    Myrmosa
Eumenidae
  Eumenes
  Euodynerus
  Pseudepipona
  Odynerus
  Gymnomerus
  Microdynerus
  Ancistrocerus
  Symmorphus

APOIDEA
Sphecidae
  Sphecinae
    Podalonia
    Ammophila
  Pemphredoninae
    Psen
    Psenulus
    Diodontus
    Pemphredon
    Passaloecus
    Stigmus
    Spilomena
  Astatinae
    Astata
    Dinetus
  Larrinae
    Tachysphex
    Miscophus
    Nitela
    Trypoxylon
  Crabroninae
    Oxybelus
    Entomognathus
    Lindenius
    Rhopalum
    Crossocerus
    Crabro
    Lestica
    Ectemnius
  Nyssoninae
    Mellinus
    Alysson
    Nysson
    Argogorytes
    Gorytes
  Philanthinae
    Philanthus
    Cerceris

Alphabetical checklist of species and their distributions

The list gives the page on which each species is named in the keys and, where applicable, the plate number. The names of the wasps follow Richards (1980) except where this has been superseded. Not all books use the same names: those in this book can be related to those in other publications by the use of Richards (1978) or Richards (1980) which also give authorities, but synonymy established since the first edition of 'Solitary Wasps' is supplied below. At first mention in a publication, the wasp's name is normally given in full, with the authority, like this: *Cerceris arenaria* (Linnaeus). *Cerceris* is the name of the genus; *Cerceris arenaria* is the name of the species; and Linnaeus is the authority who first described the species. An asterisk (*) indicates species listed by Richards (1980) but omitted from our keys.

As mentioned on p. 5, the distribution of many British wasps is poorly known. The simplified statements given here are based on Richards (1980), up-dated by reference to Falk (1991), to entomological journals and the newsletter of the Bees, Wasps and Ants Recording Scheme. Territories are indicated by abbreviations without full-stops, thus:

BI: British Isles   I: Ireland
GB: Great Britain   W: Wales
E: England   S: Scotland

Compass directions are abbreviated with a full-stop. Limits of range are given inclusively by county or, where possible, by the Watsonian vice-county, omitting 'shire' wherever permissible. A list and map of these can be found in, for example, ITE Overlays (1978).

Alysson
  lunicornis E (W. to Dorset, N. to
    Cambridge. Rare)   36, 40, pl. 7
Ammophila
  pubescens E (W. to Dorset,
    N. to S. Lancashire and Isle of Man)
    38, pl. 7
  sabulosa BI (N. to Wigtown and Moray), I
    38
Ancistrocerus
  antilope GB (N. to Ayr. Rare)   32
  gazella E, W (N. to Lancashire and N.E.
    York)   32
  nigricornis GB (N. to Inverness) 31, 32, pl. 3
  oviventris BI (commoner in N. and W.)
    32, pl. 8
  parietinus BI   32
  parietum BI (N. to S. Scotland)   32
  quadratus E (Central and S.W. Rare)   32
  scoticus BI (commoner in N. and W.,
    and by sea)   33
  trifasciatus BI (N. to Inner Hebrides)   33
Argogorytes
  fargeii E, W (W. to N. Somerset, N. to N.E.
    York. Rare)   40
  mystaceus BI (N. to S. Scotland)   40, pl. 2
Astata
  boops E, W (W. to Pembroke, N. to Norfolk)
    38, pl. 7
  pinguis BI (N. to Perth)   38, pl. 1

Cerceris
  arenaria E, W (N. to S.W. York)   39, 40, pl. 8
  quadricincta E (Kent, Essex. Very rare) 39, 40
  quinquefasciata E (W. to Devon, N. to
    Cambridge. Rare)   39
  ruficornis E (N. to W. Norfolk, before 1900
    to S. Lincoln)   39
  rybyensis E (W. to Devon, N. to Norfolk)
    39, pl. 3
  sabulosa E (one specimen, Kent, 1860)   39
Crabro
  cribrarius GB (N. to Nairn)   50, pl. 4
  peltarius BI (W. to eastern Ireland, N. to
    Sutherland)   50
  scutellatus E (W. to Dorset,
    N. to W. Norfolk)   50
Crossocerus
  annulipes GB (N. to Inverness)   54
  binotatus GB (W. to Devon, N. to
    Dunbarton)   53
  capitosus BI (N. to E. Lothian)   54
  cetratus E, W (N. to Cumberland), I   54
  dimidiatus BI (N. to Ross)   53
  distinguendus E (W. to Surrey,
    N. to S. Essex. First found 1978)   55, 56
  elongatulus E, W, I   55, 56, pl. 6
  exiguus E (W. Kent, Surrey,
    W. Sussex. Rare)   55
  leucostoma GB (S. to Isle of Man and N.E.
    York, N. to Inverness. Rare)   53

megacephalus BI   54, pl. 6
nigritus E, W (N. to N.E. York)   54
ovalis GB (W. to Devon, N. to Perth)   55, 56
palmipes E, S (W. to Devon,
   N. to Dumfries)   54, 55
podagricus BI (W. to eastern I, N. to Ayr) 53
pusillus BI   55
quadrimaculatus E, W (N. to Mid-west
   and N.E. York), I   51, pl. 4
styrius BI (N. to Dumfries.
   Rare and local)   54
tarsatus BI   55
vagabundus E (formerly W. to Hampshire,
   N. to Cheshire, now almost extinct)
                                           53, pl. 8
walkeri BI (N. to Aberdeen)   53
wesmaeli BI   55, 56, pl. 6
Dinetus
   pictus E (Berkshire, extinct)   42, pl. 1
Diodontus
   insidiosus E (W. to Dorset,
      N. to Berkshire. Rare)   45
   luperus E (W. to Devon, N. to S.E. York) 45
   minutus GB (W. to Devon, N. to Ayr)
                                           45, pl. 5
   tristis E, W (N. to N.E. York)   45
Ectemnius
   borealis E (W. Sussex, S. Hampshire.
      First found 1938, next 1972)   51
   cavifrons BI (N. to the Lothians)   52
   cephalotes E, W (N. to York and Glamorgan)
                                              52
   continuus BI (N. to Perth)   52, pl. 4
   dives E (W. to Devon, N. to Cheshire.
      First found 1972)   51
   lapidarius BI (N. to S. Scotland)   52, pl. 4
   lituratus E (W. to Devon, N. to S.W. York)
                                              52
   rubicola E, W (W. to Glamorgan,
      N. to Cambridge)   52
   ruficornis E, W (N. to N.E. York) I (Antrim).
      (Formerly rare, now increasing in
      N. England)   52
   sexcinctus E, W (N. to S.W. York)   52
Entomognathus
   brevis E, W (N. to York)   50, pl. 6
Eumenes
   coarctatus E, W (W. to Dorset,
      N. to Glamorgan. Decreasing)   30, pl. 8
Euodynerus
   quadrifasciatus E (Surrey to S. Devon.
      Very rare, mainly coastal)   31
Gorytes
   bicinctus E (W. to Cornwall,
      N. to E. Norfolk. Rather rare)   40, pl. 8
   laticinctus E (W. to Devon, N. to E. and
      W. Norfolk. Rare)   41
   quadrifasciatus E (W. to Somerset,
      N. to N.E. York)   41, pl. 2
   tumidus BI (N. to Nairn)   38, pl. 2
Gymnomerus
   laevipes E (W. to Somerset,
      N. to Leicester)   30

Lestica
   clypeata E (Surrey, 1848–53)   50, 51, pl. 8
Lindenius
   albilabris BI (N. to Nairn)   50, pl. 6
   panzeri E (W. to Dorset, N. to Norfolk) 50
   *pygmaeus E (two specimens recorded
      but may be *L. panzeri*)
Mellinus
   arvensis BI   41, pl. 3
   crabroneus E, W (formerly N. to
      Cumberland and N. Northumberland.
      Probably now extinct)   41
Methocha
   ichneumonides E, W (N. to N.E. York)
                                          29, 43, pl. 7
Microdynerus
   exilis E (W. to Dorset,
      N. to W. Norfolk. First found 1937)   31
Miscophus
   ater E (E. Kent, E. Sussex. By sea)   42
   concolor E (W. to Dorset,
      N. to Nottingham)   42, pl. 1
Mutilla
   europaea E (N. to Durham),
      S (Angus, S. Aberdeen)   29, 36, pl. 1
Myrmosa
   atra BI (N. to Cumberland)   29, 34, pl. 1, 5
Nitela
   borealis E (E. Kent, E. Sussex, Hampshire.
      First found 1982)   49
   species (reported incorrectly as *N. spinolae*;
      correct name still under investigation)
      E (Surrey, S. Hampshire. First found 1982)
                                              49
Nysson
   dimidiatus E, W (N. to Northumberland)
                                          36, pl. 2
   interruptus E (Essex to Oxford and
      Cornwall. Rare and declining)   40
   spinosus BI (N. to Mid Perth)   40, pl. 2
   trimaculatus E (W. to Cornwall,
      N. to S.E. York)   40
Odynerus
   melanocephalus E, W (W. to Devon,
      N. to Hereford and Norfolk. Rare) 30, pl. 3
   reniformis E (Surrey, S. Hampshire.
      Extinct, last record 1922)   30
   simillimus E (N. Essex, extinct; E. Norfolk,
      found 1986. Very rare; coastal marshes) 30
   spinipes BI (N. to Roxburgh)   30
Oxybelus
   argentatus E, W (N. to S. Lancashire;
      mainly coastal)   49
   mandibularis E, W (N. to Cheshire)   49
   uniglumis BI (N. to Ayr)   49, pl. 3
Passaloecus
   clypealis E (W. and N. to Cambridge.
      Rare; associated with reeds)   47
   corniger E, W (N. to York)   46, pl. 6
   eremita E (W. to S. Wilts.,
      N. to W. Norfolk. First found 1978)   46
   gracilis E, W (N. to Mid-west York)   46

insignis E, W (W. to Devon,
  N. to Mid-west and N.E. York)   47
monilicornis GB (S. to Cheshire), I   47
singularis GB (N. to Dumfries)   47
turionum E (Kent, E. Sussex,
  Surrey. First found 1987)   46
Pemphredon
  austriaca (*P. enslini*) (W. to S. Devon, N. to
  N.W. York. Very rare; nests in oak-galls)
                              43, 44
  [*clypealis* (now included in *P. morio*)]
  enslini (= *P. austriaca*)
  inornata BI (N. to Ayr)   43, 44
  lethifera BI (N. to Dunbarton)   44
  lugubris BI (N. to Midlothian)   43, pl. 5
  morio E, W (W. to E. Cornwall,
  N. to N.E. York)   43
  [*mortifer* (now included in *P. rugifera*)]
  rugifera (the current name for *P. mortifer*
  merged with *P. wesmaeli*) E
  (W. to Dorset, N. to N. Lincoln),
  S (Perth, Inverness, Moray) (Rare)   44
  [*wesmaeli* (now included in *P. rugifera*)]
Philanthus
  triangulum E (W. to S. Hampshire,
  N. to Oxford and W. Norfolk)   41, pl. 8
Podalonia
  affinis E (W. to Hampshire, N. to
  Lancashire), I (Wicklow)   38
  hirsuta E, W (N. to S. Lancashire) 38, pl. 2
Polistes
  dominulus E (a rare immigrant social wasp
  which will run incorrectly to key III)
Psen
  ater E (N. Hampshire, E. Suffolk; records
  are all more than 100 years old)   34
  atratinus E (Isle of Wight.
  First found 1950)   35
  bicolor E (W. to S. Devon,
  N. to S.W. York)   37
  bruxellensis E (W. to Dorset and Glamorgan,
  N. to Glamorgan and Middlesex;
  S. Lancashire)   37
  dahlbomi GB (W. to Pembroke,
  N. to W. Inverness)   35
  equestris GB   37
  littoralis E, W (E. to Dorset,
  N. to N. Lancashire), I (east). (By sea) 35
  lutarius E (W. to Dorset,
  N. to S.W. York)   37, pl. 2
  spooneri E (W. to Dorset, N. to W. Norfolk)
                                  35
  unicolor E (W. to Isle of Wight and S.
  Hampshire, N. to S. Essex.
  First found 1950)   35
Psenulus
  concolor E (W. to Devon, N. to York) 35
  pallipes (*P. atratus*) E, W (W. to N. and S.
  Devon, N. to N. Lancashire), I (east and
  central)   35, pl. 5
  schencki E (W. to W. Sussex, N. to
  Cambridge. First found 1954)   35

Pseudepipona
  herrichii E (heaths of south-east Dorset.
  Very rare)   31
Rhopalum
  clavipes BI   49, pl. 4
  coarctatum BI (W. to Fermanagh,
  N. to Inverness)   50
  gracile E (W. Suffolk, Cambridge,
  E. Norfolk. Fens; very rare)   50
Sapyga
  clavicornis E (W. to N. Devon,
  N. to N.E. York)   41, pl. 7
  quinquepunctata E, W   36, pl. 1
Smicromyrme
  rufipes E (W. to Devon,
  N. to Norfolk)   29, 36, pl. 7
Spilomena Very small wasps; species difficult
  to identify   42, pl. 5
  *beata E (W. to Devon, N. to Stafford)
  *curruca (previously *differens*) E (W. to
  Devon, N. to Warwick)
  *enslini E (south-east), I (Queens County)
  *troglodytes E, S (W. to Devon, N. to
  Dumfries)   pl. 5
  [*vagans* (now included in *S. troglodytes*)]
Stigmus
  pendulus E (E. Kent, Middlesex, S. Essex,
  Cambridge. First found 1986)   42
  solskyi E, W (N. to York)   42
Symmorphus
  bifasciatus (*S. mutinensis*) BI (N. to Perth)
                             33, pl. 3
  connexus E (W. to Dorset, N. to N. Lincoln.
  Rare)   33
  crassicornis E (W. to Devon, N. to S.E. York.
  Rather rare)   33
  gracilis E, W (N. to N.E. York)   33
  mutinensis (= *S. bifasciatus*)
Tachysphex
  obscuripennis E (one specimen, Kent, 1882)
                              38
  pompiliformis BI (W. to eastern I, N. to
  Aberdeen)   38, pl. 7
  unicolor E, W (W. to Devon,
  N. to N. Lincoln)   34
Tiphia
  femorata E, W (N. to Norfolk)   45, pl. 5
  minuta BI (N. to Ayr)   45
Trypoxylon
  attenuatum GB (N. to Dumfries)   48
  clavicerum E, W (N. to N.E. York)   48
  figulus GB (W. to S. Devon, N. to W.
  Norfolk and perhaps to Dumfries. Now
  apparently being supplanted from the S.
  by *T. medium*)   48, pl. 6
  medium E (W. to S. Devon,
  N. to W. Norfolk)   48
  minus E (Kent. One specimen, 1959) 48

Index

CPSIA information can be obtained
at www.ICGtesting.com
Printed in the USA
BVHW020849301020
591953BV00007B/95